The City in Maps: urban mapping to 1900

The City in Maps:

James Elliot

The British Library

urban mapping to 1900

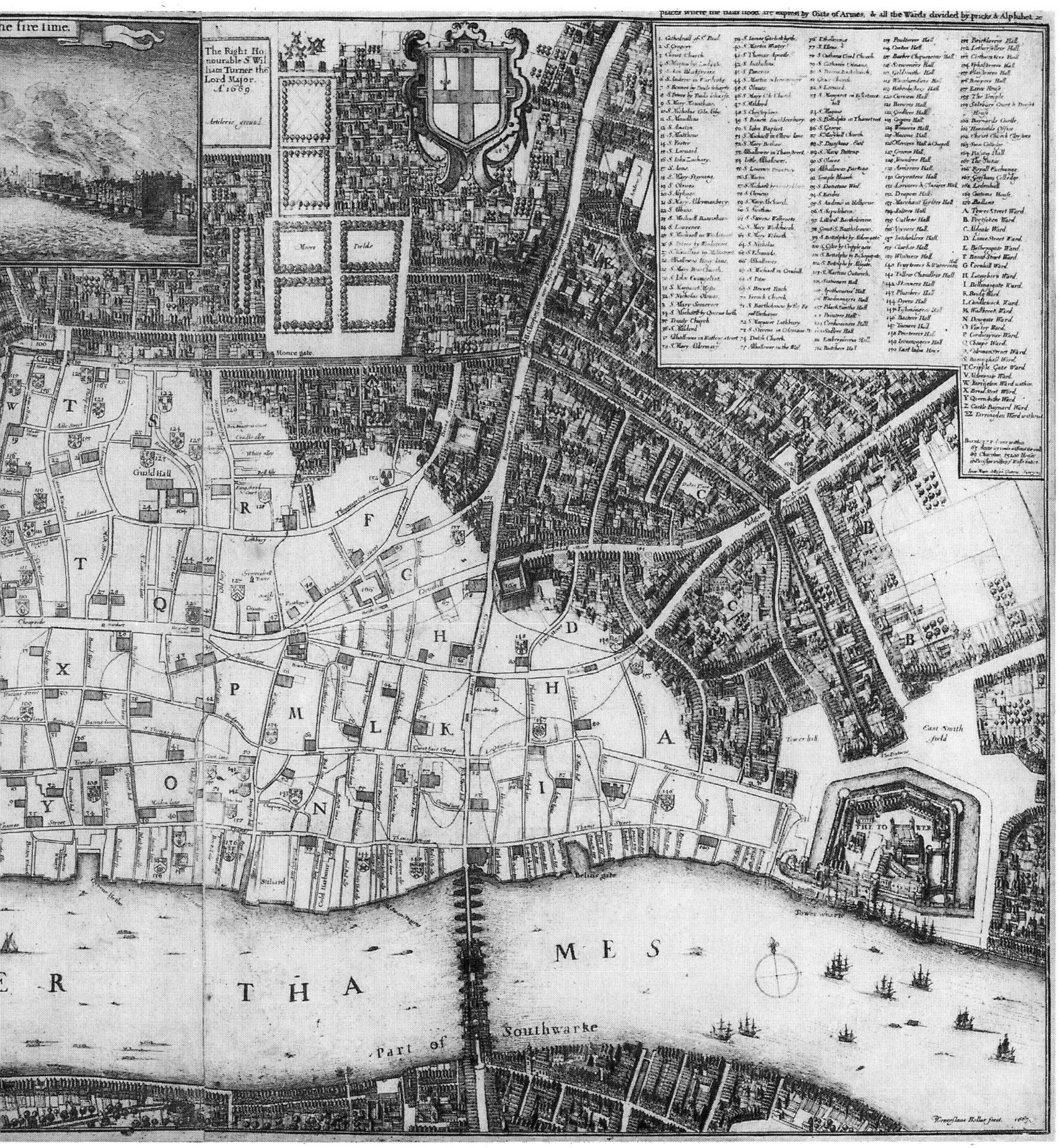

First published 1987
by The British Library
Great Russell Street
London WC1B 3DG

British Library Cataloguing in Publication Data
Elliot, James
The city in maps: urban mapping to 1900.
1. Cartography—History 2. Cities and towns—Maps
I. Title II. British Library
526′.09173′2 GA203

ISBN 0-7123-0134-8

Designed by Alan Bartram
Typeset in Monophoto Photina
Colour origination by York House Graphics, Hanwell
Printed in England by Jolly & Barber Ltd, Rugby

FRONT COVER
Anon, 'A Description of the Towne of Mannados or New Amsterdam as it was in September 1661' (the 'Duke's plan' of New York). (1664).
BL Maps K.Top. CXXI.35.
New Amsterdam was the name given to New York by the city's Dutch founders. The presence in the harbour of English ships, however, dates this plan to some time after September 1664 when the English took the town and renamed it New York in honour of James, Duke of York. Several of the landmarks of the modern city can be seen, including the fort that gave its name to the Battery Park, and the pallisaded ditch along which Wall Street runs today.

BACK COVER
Jacopo de' Barbari woodcut view of Venice. (Venice, 1500.) Detail
[See **11**.]

FRONTISPIECE
'An exact surveigh of the . . . ruines of the City of London', Wenceslaus Hollar. (London: John Leake, 1669.)
[See **45**.]

Contents

Acknowledgements

The exhibition on which this book is based was organised by the British Library Map Library. Its preparation was the work of James Elliot, under the general guidance of Dr Helen Wallis. The design of the exhibition was undertaken by John Ronayne.

The British Library Board wishes to thank the Trustees of the British Museum for the loan of the clay tablet of Tuba and de' Barbari's view of Venice.

The compiler wishes to express his warmest appreciation of the help and assistance readily given by his colleagues in the British Library for the preparation of both the exhibition and the book. The following individuals deserve particular mention: Dr Helen Wallis and Sarah Tyacke; Tony Campbell and Peter Barber, who contributed much early material and who were an invaluable source of ideas and constructive criticism; Terry Kay, who handled the administration; Alan Sterenberg, the acting exhibitions officer; Anne de Lara, Angela Roche and Shelley Jones, who mounted the exhibition.

The compiler gratefully acknowledges Ralph Hyde of the Guildhall Library, London, for his advice on the selection of exhibits.

Lastly, the compiler would like to thank his wife, Sue, for her patience and support throughout the project.

Preface

This is, I believe, the first time that the Map Library of the British Library (or the Map Room of the British Museum, as we were before 1973) has put on an exhibition to illustrate the mapping of cities through history. In choosing this theme for our Map Gallery display, we were prompted not by any particular anniversary but simply by the wish to show off some of our treasures and significant material in the field of city maps.

The subject is an enthralling one. The city has featured as the hub of civilization in all those cultures where people have established a settled life organised around central places. Cities have evolved a diversity of functions, from ceremonial and religious centres to fiscal and administrative control points and emporia of trade. They serve as powerful symbols of the civilizations whence they derive. Sennacherib's account of the destruction of Babylon, 'the grandeur that was Rome', visions of Jerusalem the golden and of New Jerusalem, 'London . . . the flower of cities all!': these inscriptions and sayings conjure up visions of great cities and the rise and fall of empires. The maps themselves reflect the variety of functions and the prestige of the city. Urban mapping has called out the highest skills of the surveyor, plate-maker, artist and engraver.

This is the last Map Library exhibition prepared under my aegis before my retirement as Map Librarian. With special pleasure therefore I pay tribute to James Elliot, the organizer of the exhibition and author of this book. He has selected a wide range of items from the holdings of the Map Library, which include major atlases of the sixteenth and seventeenth centuries and King George III's Topographical Collection, perhaps the finest eighteenth-century map collection in the world. He has supplemented these with material from other departments of the library. Loans from the British Museum comprise a Babylonian tablet of the 4th century BC and de' Barbari's plan of Venice. The result is a multicultural display that embodies many aspects of the development of urban civilization, historical, aesthetic and psychological, over a time-span of some 2,500 years. To the enthusiasm and expert researches of James Elliot and his team of collaborators we are indebted for a book which can be enjoyed by a wide audience as a permanent record of a fascinating exhibition.

HELEN WALLIS, November 1986

(1) Fragment of a city plan labelled 'Tuba', drawn on clay (4th century BC).
British Museum. Department of Western Asiatic Antiquities, no. 35385.
The plan, with south-east at the top, shows the River Euphrates, identified by the water-lined band. In the centre is Tuba, below which are part of the city walls and the Shamash Gate, or 'Great Gate of the Sun God'. On the reverse of the tablet, cuneiform text describes the defence measures for the city, including the closure of the gates after dark and the posting of sentries.

Introduction
The earliest town plans

To some kind of men it is an extraordinary delight to study, to looke upon a geographicall map and to behold, as it were, all the remote Provinces, Towns, Citties of the world . . . what greater pleasure can there be then . . . To peruse those books of Citties, put out by Braunus and Hogenbergius.
Robert Burton, *The Anatomy of Melancholy*, 1621.

The modern user of the town plan is familiar with a clear and functional publication which conveys a range of information limited to the accurate representation of the spatial features of the urban landscape around him. These features, moreover, are symbolised and so reduced to a basic cartographic code which is designed to transmit in the most consistent and simple form possible the topography of the surrounding streets, the whereabouts of the nearest hospital, the most direct route to the nearest railway station, and a variety of other soundly prosaic messages. The modern plan represents a strictly two-dimensional attempt to preserve the correct spatial relationships between streets and buildings by maintaining throughout a uniform scale.

To Robert Burton and his seventeenth-century contemporaries, however, town plans evidently offered more, being a source of 'extraordinary delight'. The reason for this was that by Burton's time urban cartographers were often concerned not only with showing the street layout, but also with depicting the architectural splendours of the cities they knew. Map-makers, indeed, frequently sacrificed the precision of a ground plan for the sake of more pictorial cartographic styles. These offered the flexibility necessary to convey not only all three dimensions of the urban form but also some of the imaginative and symbolic qualities of a work of art, rather than a scientific document dependent for its creation purely on survey and measurement.

One such style was the bird's-eye view, in which the city was depicted from a high oblique angle similar in effect to that which might be obtained from a balloon or aeroplane. The bird's-eye view enabled the cartographer to convey the vertical dimension of the buildings and architectural features of the city, while at the same time retaining a horizontal dimension, albeit one which relied on perspective instead of true scale. The plan and the bird's-eye view could be combined to form the map-view or plan-view, in which the true ground plan was preserved, but which featured some or all of the buildings in elevation. By depicting a city in these ways, the cartographer's intention was clearly to impress and inspire the reader with the grandeur, power and wealth his works displayed. Urban cartography thus possessed a quality which sought a fundamentally emotional response, one which reflected the pride, dignity and sense of importance the city-dweller felt for his community.

(**2**) Fragment of a large plan of the city of Rome (the 'Forma urbis Romae') executed on stone, 203–11 AD. *From* Jordan, H. *Forma urbis Regionum XIIII.* (Berlin, 1874) BL 7705.i.3.

It is clear that the oldest urban civilisations found some purpose in the cartographic representations of their settlements. The earliest depiction extant dates from *c*.6200 BC, and is in the form of a wall painting of Çatal Hüyük, central Turkey. The painting shows, in stylised form, the houses and main thoroughfare of the town, dominated in the background by a nearby volcano. The earliest ground plans, however, came from the sophisticated urban culture of Mesopotamia, and survive as fragments of clay tablets on which the plans were incised. One of these fragments, dating from the fourth century BC, shows a place known as Tuba (**1**), a suburb of the city of Babylon. It demonstrates the typically Babylonian style of incising in straight strokes where possible, as the clay was usually too soft to permit curved lines.

Although no town plans are known from ancient Greece, the deliberately symmetrical form of cities such as Megalopolis (founded 371 BC) indicates that urban surveying was a well-developed science. The Roman Empire, always an urbanising civilisation, required the skills of trained *agrimensores* (land surveyors), whose most impressive work survives in the form of 679 fragments of marble tablets, bearing part of a large plan of Rome (**2**). Known as the 'Forma urbis Romae', it was executed in 203–11 AD on a scale of about twenty feet to the inch (1:240).

After the end of the western empire in the fifth century AD, which brought the breakdown of urban life, the techniques of the *agrimensores* in producing the measured town plan became lost to Europe. In China, by contrast, an unbroken tradition of surveying and urban mapping continued from at least the third century BC to the time of the Yuan Dynasty (1280–1368), as attested by the survival of several town plans, including one of the city of Suchow carved on stone in 1229 AD.

I

The medieval city

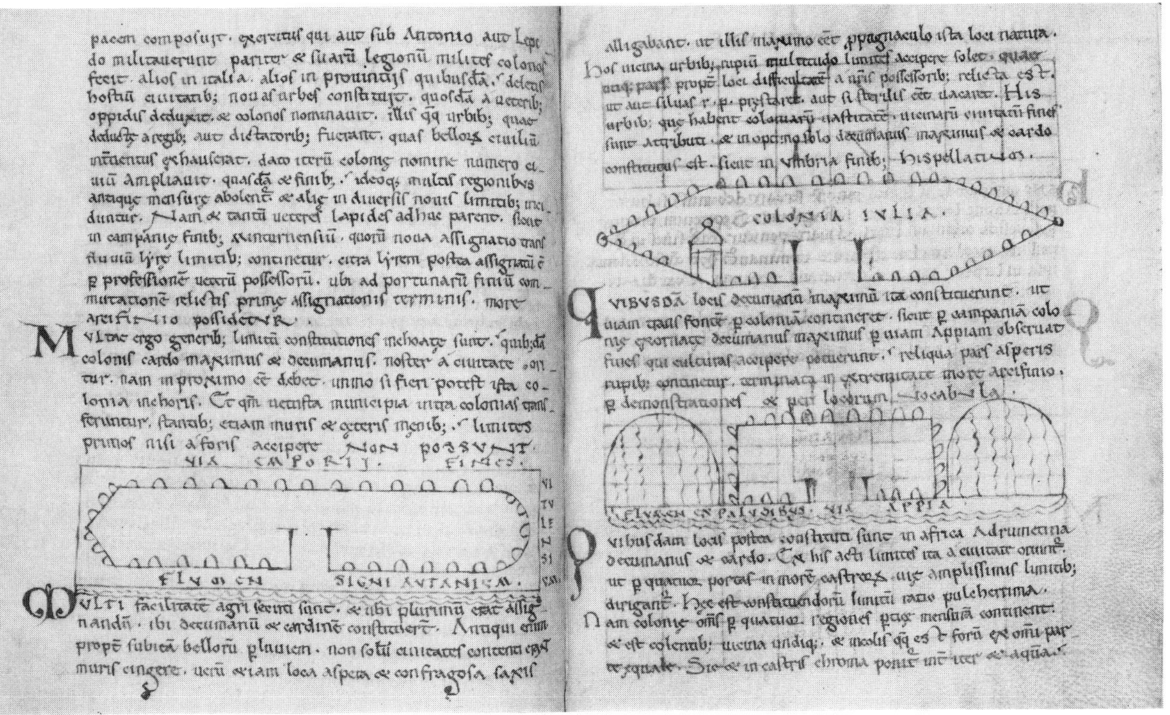

The tradition of the measured survey established in Roman times re-emerged in Europe between the ninth and twelfth centuries with the copying of the treatises of the *agrimensores*. They not only produced precisely-measured plans but also used unmeasured pictorial and diagrammatic maps of towns and their environs to illustrate their surveying texts, known collectively as the *Corpus agrimensorum*. One such example (3), a medieval copy of a work attributed originally to Hyginus Gromaticus (fl. early second century), considers the problems of dividing up lands to form *coloniae*, or settlements for deserving military veterans. The text outlines the difficulties encountered when the surrounding topography was not suitable for the precisely-measured grids upon which the *coloniae*, and the towns which formed their nucleii, were planned. When the Roman text was compiled in 350–400 AD, it was illustrated with crude copies of original working plans of about 50 AD. These were again copied, and further corrupted, in the twelfth century in western Germany or France, and survive as unscaled 'picture maps', in which walls and buildings were drawn in profile on a simplified

(3) 12th-century copy, executed in western Germany or France, of a Roman text on estate surveying ascribed to Hyginus Gromaticus (*fl.* early 2nd century) with illustrations originally of *c.*50 AD. BL Additional MS 47679, ff.63v–64. This text is one of a number of works known collectively as the 'Corpus Agrimensorum'. On the left page is an unidentified plan of a town between a river and a principal road. On the right is a diagram of the walled town of 'Colonia Julia' or Hispellum (now Spello) in Umbria. The lower illustration shows the walled town of Tarracina (now Terracina) south of Rome, then surrounded by a river and the Pontine Marshes ('Flumen en Paludibus').

ground plan. These final copies were made at a time when renewed interest was being taken in surveying, possibly in connection with the creation of new towns to meet rapid population growth.

The 'books of islands', or *isolarii*, originally compiled as navigational aids for Mediterranean sailors, were also illustrated by picture-maps (PLATE 1). By 1420, when Cristoforo Buondelmonte's 'Liber Insularum Archipelago' first appeared, these *isolarii* had evolved into sophisticated travel guides containing not only maps and plans but accounts of the history, legends and geography of the islands and harbours in the region. The 'Liber Insularum Archipelago' became one of the most popular books of the time and was frequently imitated, with copyists adding to the maps or interpreting them afresh. One of the finest surviving examples of Buondelmonte's work is the 1482 Ghent version commissioned by Raphael de Marcatellis, Abbot of St Bavon. The illustrations may be by Jan van Kriekenborch of Ghent, who seems to have specialised in the production of manuscript maps. It was natural for Buondelmonte to have included Constantinople (Istanbul) since in the 1420s it was the most important city in the region and, until 1453, the capital of the Byzantine Empire. As with Matthew Paris's London (PLATE 5) the city is identified through the depiction of its major buildings, and an attempt has also been made to give an idea of the street pattern within the walls. A much later example of the 'picture-map' style can be seen in a Persian manuscript plan of the *Haram*, or Holy Sanctuary of Mecca (**4**) of the late sixteenth or early seventeenth century.

In contrast to the 'picture-map', plans of the cities of Jerusalem and Acre (**5**) by the Genoese-born Pietro Vesconte display an awareness of proportion which predates the first scale plans of Renaissance Europe by almost 200 years. They were drawn in *c.*1320–25 to illustrate the 'Liber Secretorum fidelium crucis', a propaganda tract advocating a new crusade to recover the Holy Land, written by the Venetian diplomat and traveller Marino Sanudo Torsello. Both plans were based on the accounts of crusaders and pilgrims as well as on other material which now no longer survives.

Realistic views of European towns appear regularly from 1400 onwards as backgrounds to paintings, and in manuscript miniatures, particularly in Flanders and central Italy. The first printed representations appear in the 1470s as simple woodcut prospects used for book illustration. The earliest known is a view of the city of Cologne – complete with the crane on its then unfinished cathedral – in Werner Rolewink's world chronicle *Fasciculus Temporum* of 1474. This work also contains a number of prospects that were the product of the author's imagination.

Perhaps the earliest printed book in which the topographic contents are significant for their general accuracy is the *Peregrinatio in Terram Sanctam*, 1486 (**6**). It was compiled by Bernhard von Breydenbach, a wealthy canon of the cathedral at Mainz, who journeyed to the Holy Land in 1483. He was accompanied by the Dutch artist Erhard Reuwich, whose drawings were used to prepare woodblocks of the towns visited *en route*, including Venice, Rhodes, Candia (Iraklion) and Jerusalem. These illustrations supplemented Breydenbach's text, which thus became the first illustrated travel book as well as the first book to contain folding plates. Another first-hand account containing a folding illustration was

(**4**) Muhyi Lari, leaf from a manuscript of 'Futuh al-Haramayn' (Blessings of the Two Sanctuaries), a Persian poem describing the holy places of Islam. This leaf depicts the Holy Sanctuary of Mecca. (Later sixteenth or early seventeenth century.)
BL OMPB Oriental 11533. f.18v.

باب السلام
باب النبي
باب عباس
باب علي

(**5**) Maps of Jerusalem and Acre, by Pietro (or Pierino) Vesconte. From Marino Sanudo Torsello, 'Liber Secretorum fidelium crucis', *c*.1320–25.
BL Additional MS 27376*, ff.189v–190.
The plan of Jerusalem (left) shows the biblical city rather than the medieval one. The homes ('domus') of King Solomon, Pontius Pilate, St Anne and the area occupied by the Holy Sepulchre ('sepulchrum domini') and the Temple ('Area templi') are indicated, but not the places of worship that later occupied these sites.

The sparseness of the Jerusalem plan, which had passed from Christian control some 125 years earlier, contrasts with the detail shown for Acre (right). This last bastion of the crusaders had succumbed to Islam within living memory in 1290, and it is depicted in its last days as a Christian stronghold. The fortified harbour, the arsenal, the castle ('castellum') and soldiers' lodgings (*eg* 'hospitum hospitalis') are indicated, as are the parts of the walls and town that were in the custody of the different nations (*eg* 'turris anglorum') and of the crusader orders (*eg* 'custodia templariorum').

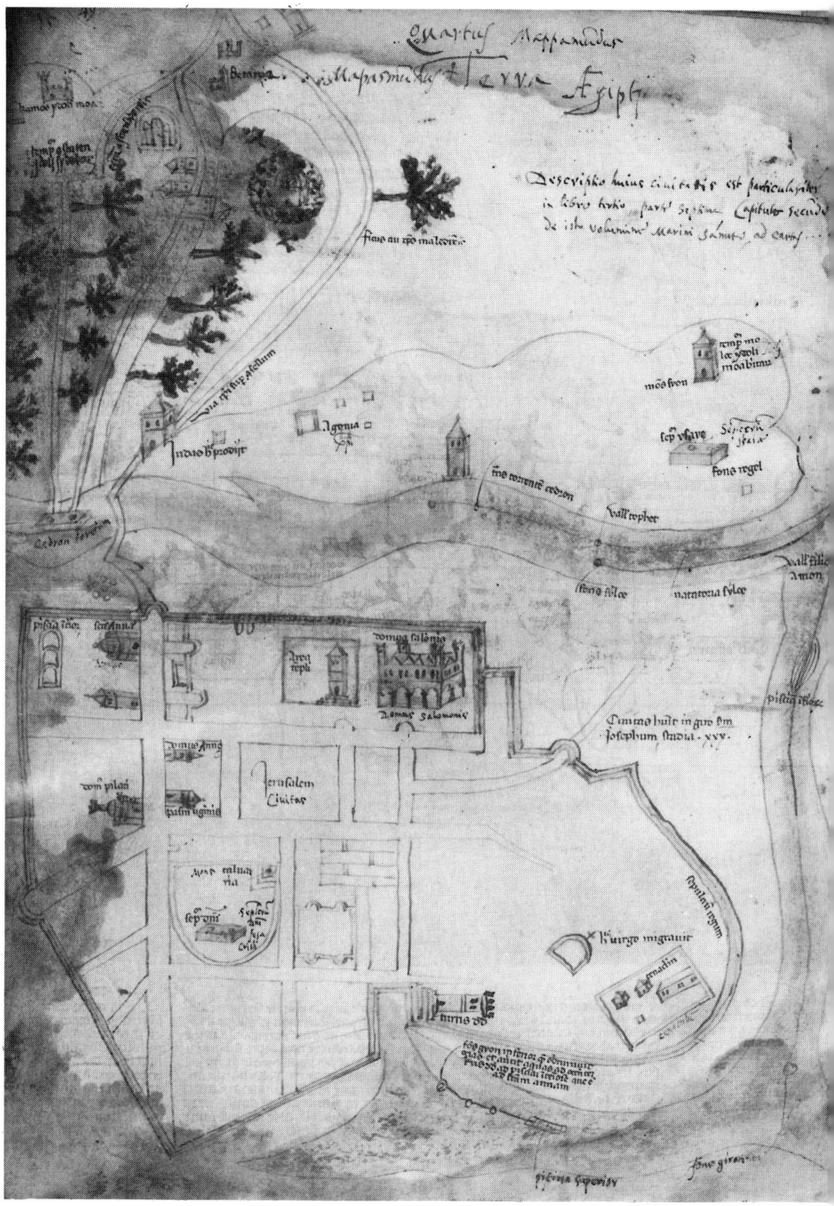

(**6**) 'Rodis'. (View of Rhodes by Erhard Reuwich). From Bernhard von Breydenbach, *Peregrinatio in Terram Sanctam* . . . (Mainz, Erhard Reuwich, 11 February 1486.)
BL C.20.e.3.
The view shows the effects of the terrific Turkish siege of 1480, three years previously. The Tower of St Nicholas (labelled in the foreground) had been partly destroyed, but by the time of Breydenbach's visit was largely rebuilt. Behind the tower, Reuwich shows a ship being caulked, and in the Greek Harbour, to the left, the galley that had brought the party.

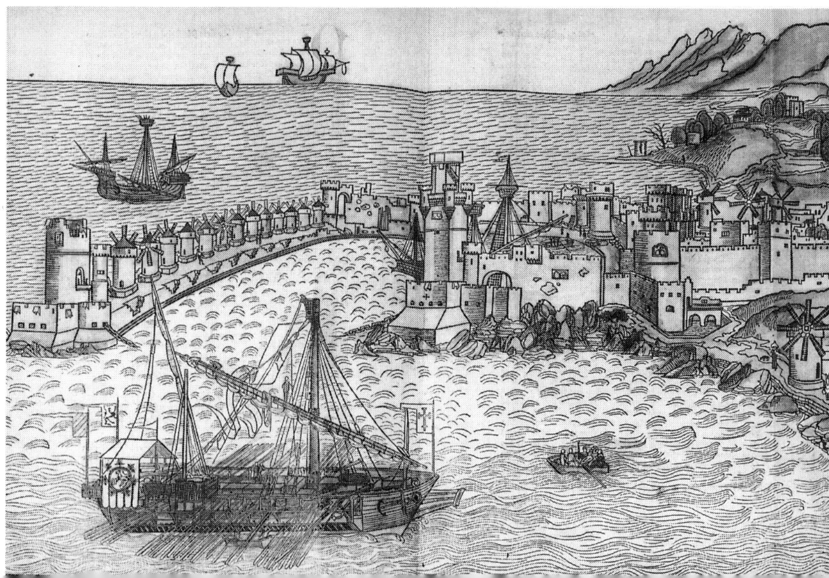

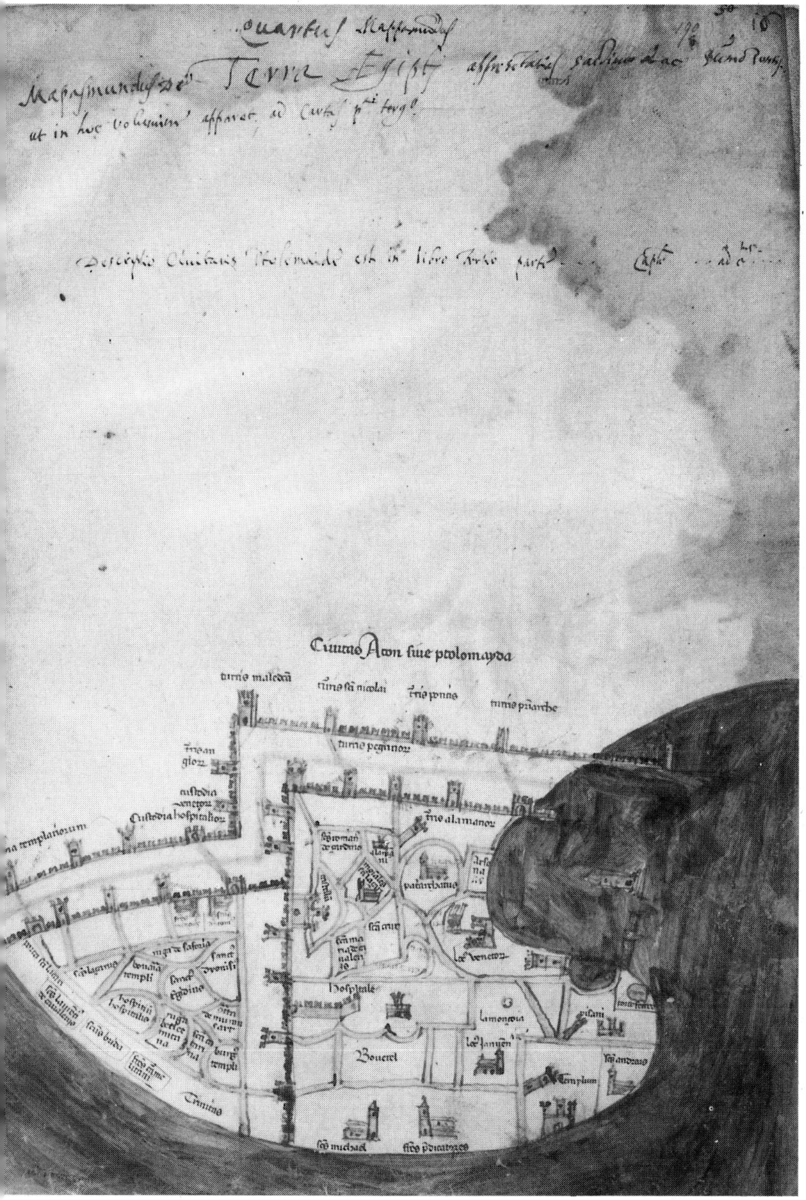

Ciuitas Acon siue ptolomayda

(7) Plan-view of Mexico City and separate chart of the Gulf. From Hernando Cortés, *La preclara narratione di Ferdinando Cortese della Nuoua Hispagna*. (Venice: Bernardino de Viano for Baptista de Pederzani, 1524.) BL 9771.b.11.
The centre of this 'plan view' is the great central square of the Aztec capital Tenochtitlan (here Temixtitan), bordered on one side by Montezuma's palace and complete with rows of sacrificial heads. Also discernible are the causeways linking the island city to the shores of Lake Texcoco.

the second of six letters written by Hernando Cortés giving his description of the conquest of the Aztec empire in 1521. The 1524 Venice edition contains a woodcut plan-view of the Aztec capital Tenochtitlan (7) which is probably the earliest printed plan of any American city. By 1524, however, the plan had become a valuable historical record, as Cortés had destroyed the original capital to make way for his new foundation, Mexico City.

The projects of Breydenbach and Cortés were exceptional in the overall accuracy of the plans and views they contained. Generalised views, taking the form of largely imaginary sketches in which the same woodblock would frequently be used to represent several towns, remained the favoured method of depicting towns until well into the sixteenth century. The *Nuremberg Chronicle* of 1493, for example, contained

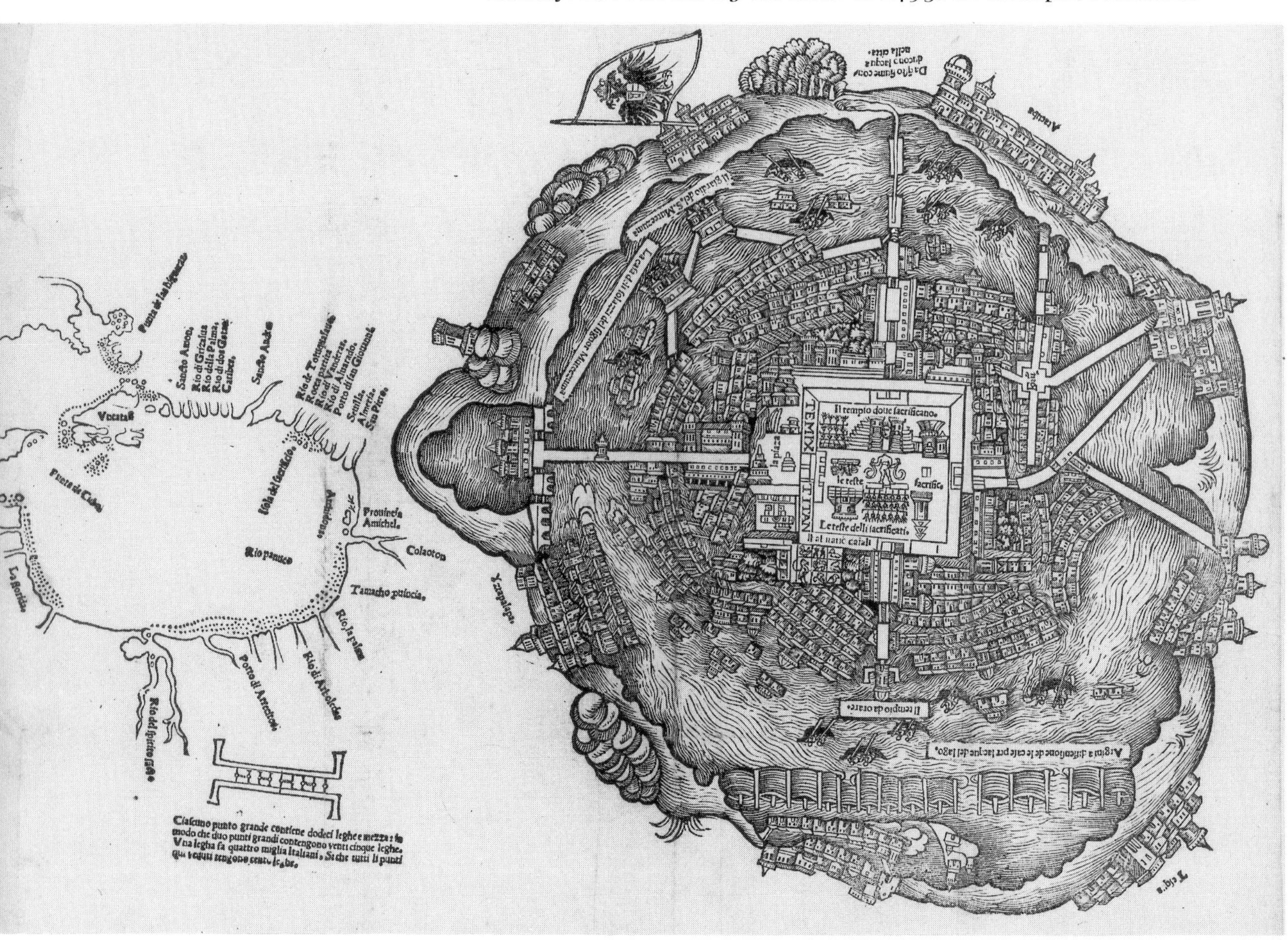

(PLATE 1) Depiction of Constantinople. From Cristoforo Buondelmonte, 'Liber Insularum Archipelago' (1420) as copied in Ghent or Bruges in 1482 for Raphael de Marcatellis, Abbot of St Bavon, Ghent.
BL Arundel MS 93, ff.154v–155.
Prominent among the buildings shown are the domed St Sophia (Hagia Sophia) prior to its conversion into a mosque, the Imperial Palace (top left), the Hippodrome ('ypodromos') in front of St Sophia and (bottom left) the church of St John of Studius ('sanctus Johannes de studio'). The residential quarter of Pera can be seen across the great harbour known as the Golden Horn, while on the Asian side of the Bosporous is the town of Scutari.

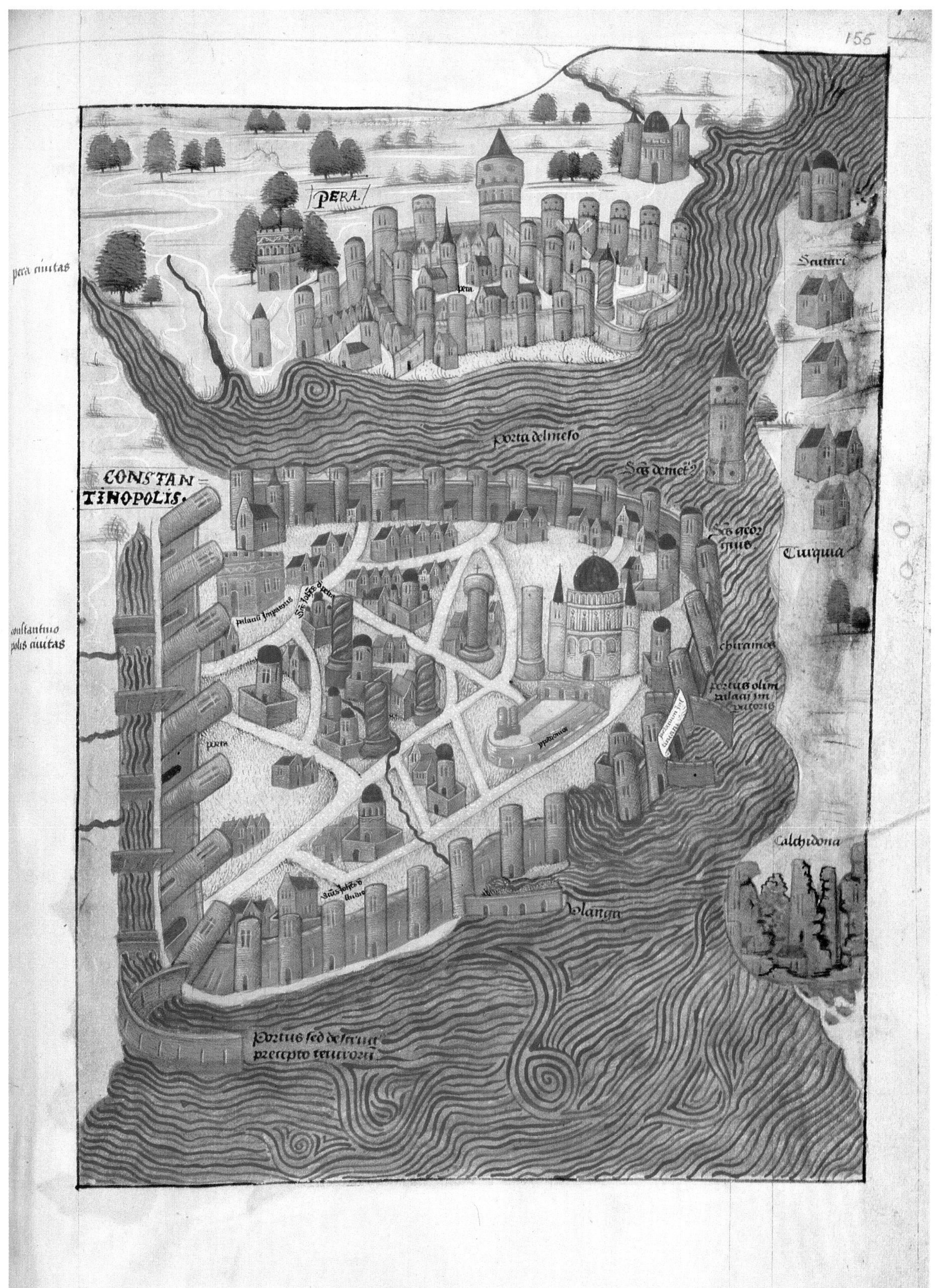
155
PERA
pera ciuitas
Porta del meso
CONSTAN
TINOPOLIS
constantino
polis ciuitas
Turquia
Calchidona
Vlanga
Portus sed destruct
precepto teucrorum

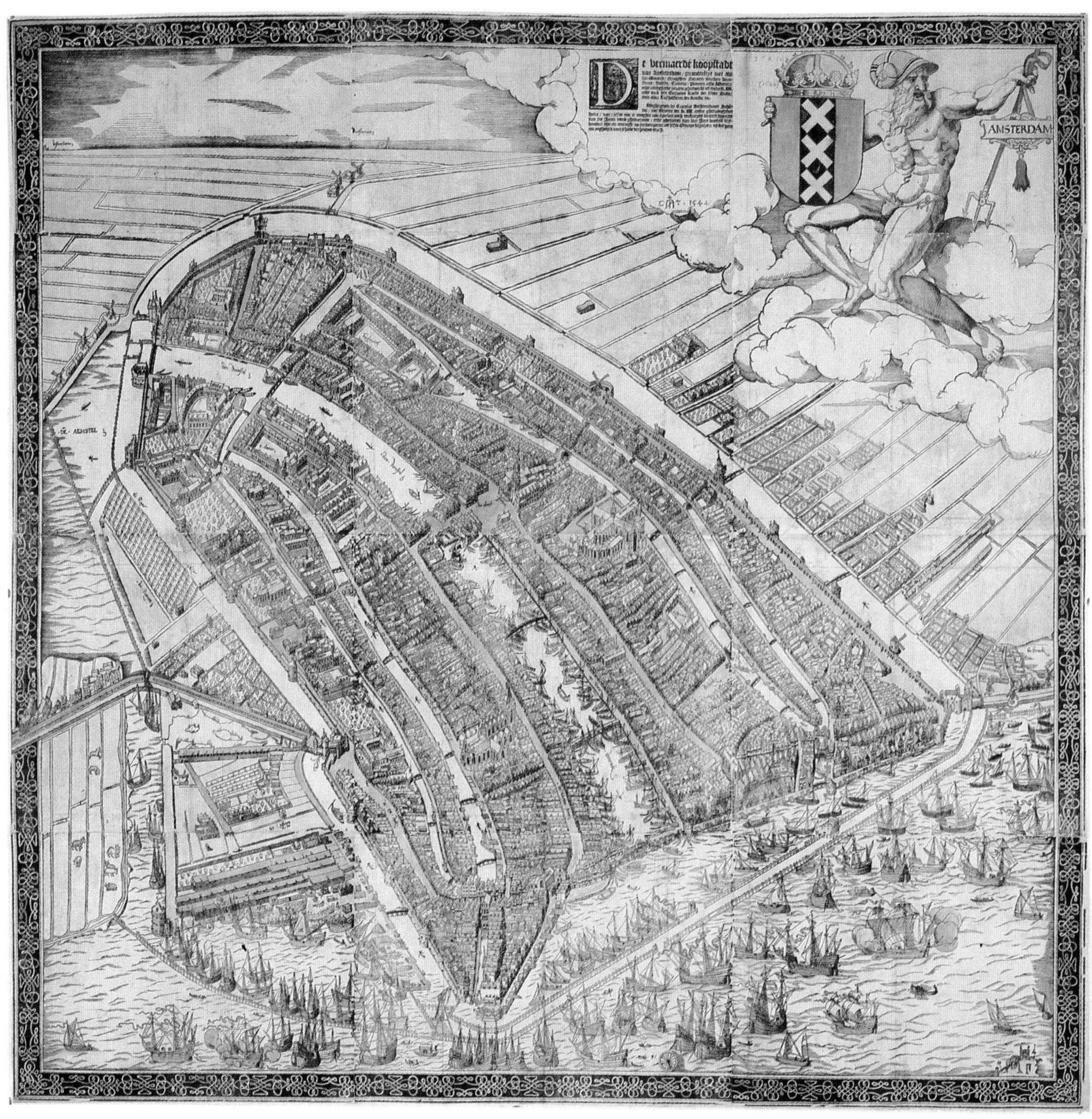

(PLATE 2) Cornelis Anthoniszoon, 'De vermaerde Koopstadt van Amstelredam'. (Amsterdam, 1544.)
BL Maps S.T.A.(4).

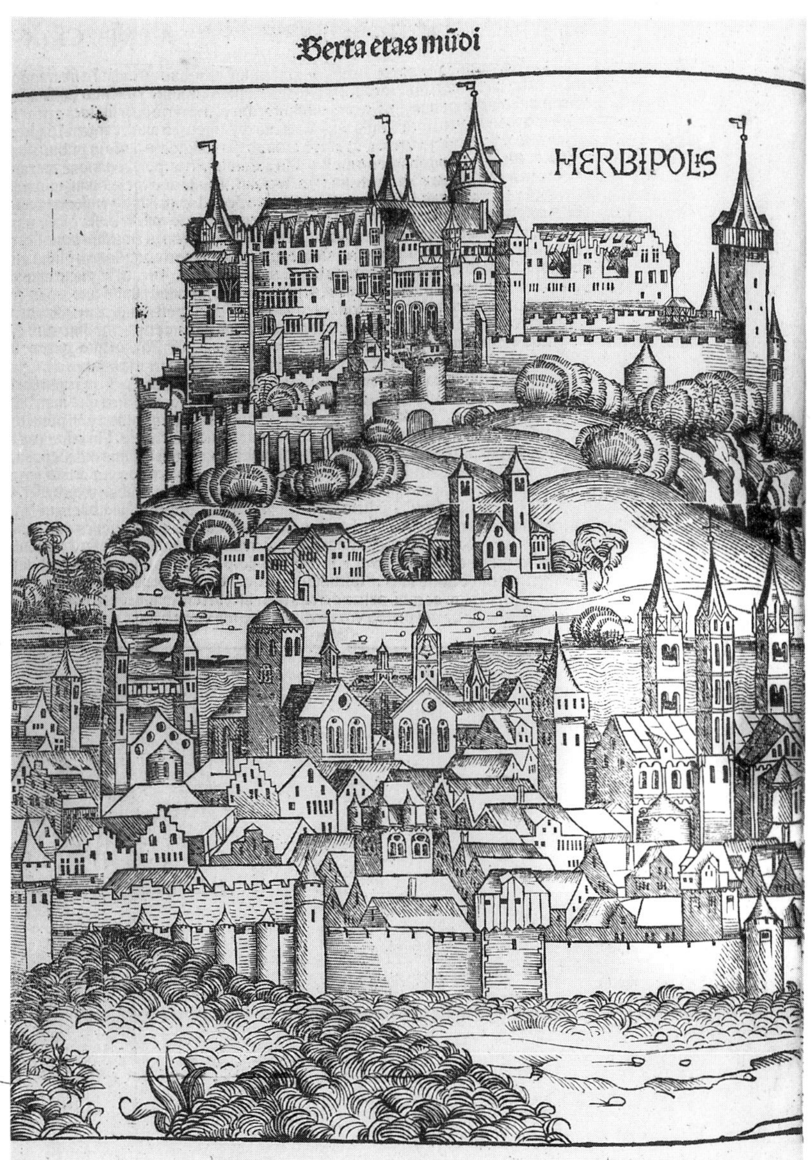

(**8**) View of Wurzburg. From Hartmann Schedel, *Liber cronicarum*. (Nuremberg, 1493.) BL IC.7432.

116 named views of places, of which perhaps less than a quarter approximate to the actual scene (**8**). Sebastian Münster also used a few woodblocks to portray a large number of towns in the first edition of his *Cosmographia* of 1544 (**9**). The *Cosmographia* was one of the most influential books of its day, a geograpical and historical compendium drawn from many classical and contemporary sources, from the leading humanists of the time, and supplemented by Münster's own observations during his travels. The publication in 1548 of Johnanes Stumpf's *Schweizer Chronik* (**10**), however, was notable as it carried the first realistic views of Swiss cities, the accuracy of which represented a great advance in geographical illustration. This stimulated Münster to revise his own work and to include substantial improvements in both the quantity and quality of the town views in the 1550 edition of the *Cosmographia*, thus providing Renaissance scholars with their first comprehensive view of urban Europe.

(9) View of Metz. From Sebastian Münster, *Cosmographia. Beschreibung aller Lender*. (Basle: Heinrich Petri, 1544.)
BL Maps C.8.a14, p.lx.
A generalised view, complete with nesting stork, used by Münster to represent several towns including Metz and Arles. [Same size as original.]

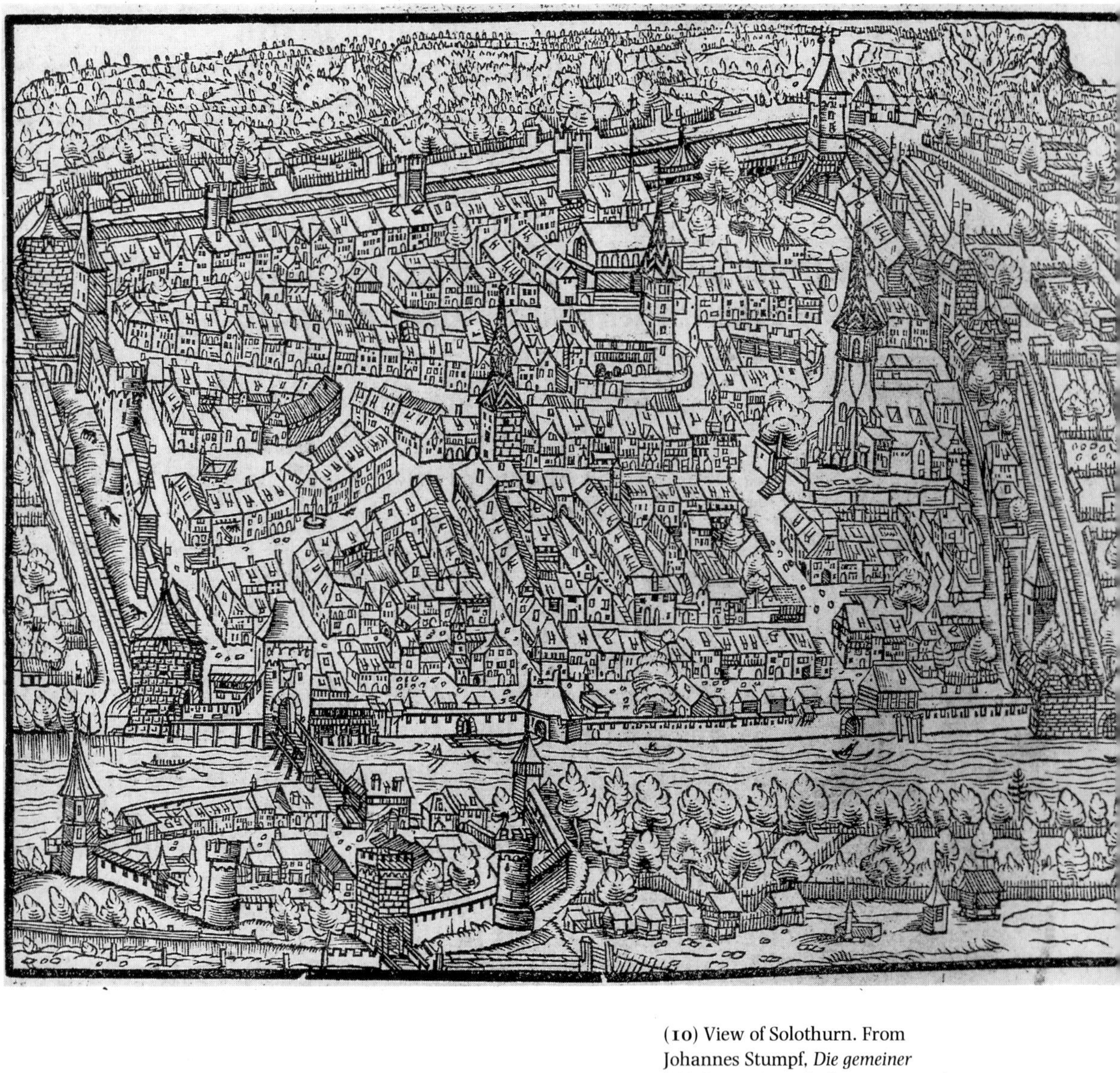

(10) View of Solothurn. From Johannes Stumpf, *Die gemeiner loblicher Eydgnoschafft Stetten, Landen und Volckeren Chronik wirdiger thaaten beschreybung* (the 'Schweizer Chronik'). (Zurich: Christoph Froschauer, 1548.)
BL 1482.g.9, Volume 2, f.223.

The European Renaissance witnessed developments in aspects of mathematics and the graphic arts that were to have a direct impact on urban cartography. The science of perspective, advanced through Brunelleschi's projective geometry of 1425, and Leon Battista Alberti's method of perspective drawing set out in 1435, led in Italy to the development of the bird's-eye view. This offered a method of achieving a realistic impression of an urban area from an otherwise unattainable viewpoint. Several such views of Italian cities, including those of Florence and Rome, are known to have been produced in the 1480s and 1490s. The most spectacular, however, was a view of Venice published in 1500 (11). This work was originally attributed to Albrecht Dürer, and not identified until 1855 as the work of Jacopo de' Barbari. An itinerant painter born in Venice *c.*1450, de' Barbari was court artist to a number of princes, including the Emperor Maximilian I. The view of Venice was commissioned by a German merchant resident in the city, Anton Kolb, who approached the Venetian government in 1500 for a grant of the sole publishing rights and an exemption from export duties. The view was probably composed in a studio from many separate sketches taken from towers and other high buildings. The individual sketches would then have been plotted onto a perspective framework which itself may have been calculated from ground plans of the city similar to those used by the Venetian authorities for the maintenance of roads and canals. The design was cut on six woodblocks which still survive in the Correr Museum, Venice. De' Barbari's work epitomised the new heights of realism to which the Renaissance cartographers aspired. At the same time, the symbolic purpose of his work can be deduced from the Latin inscriptions encircling Mercury, hovering above the city, and Neptune, riding a dolphin below. These may be rendered as 'I, Mercury shine favourably on this above all other emporia', and 'I, Neptune, reside here, smoothing the waters of this port'.

De' Barbari's view of Venice was probably the inspiration for the woodcut view of Amsterdam executed in 1544 by Cornelis Anthoniszoon (PLATE 2). This again features Neptune, here positioned high above the city, bearing a trident in his left hand and Amsterdam's coat of arms in his right. Anthoniszoon, a topographical illustrator, was of sufficient stature to be taken by the Emperor Charles V on his expedition to Algiers in 1541 to provide a pictorial record. His view of Amsterdam, of which five different editions are known to have been published between 1544 and *c.*1640, mirrored the pride he felt for his native city as suggested by the title, 'De vermaerde Koopstadt van Amstelredam' (the famous trading city of Amsterdam).

By this time, the bird's-eye view had evolved as a popular and familiar means of portrayal. A more specialised application of the technique can be seen in the views of battles and sieges produced from about 1490. Antonio Lafreri of Rome published numerous depictions of military campaigns from the mid 1560s onwards. His work also included 'Le sette chiese di Roma' (the seven churches of Rome), which he made for the 1575 papal jubilee (**12**). His engraving shows pilgrims processing between the seven churches on the route, with all other details – apart from the Tiber and the city walls – excluded.

Italy also led the way in the science of fortification and siege warfare;

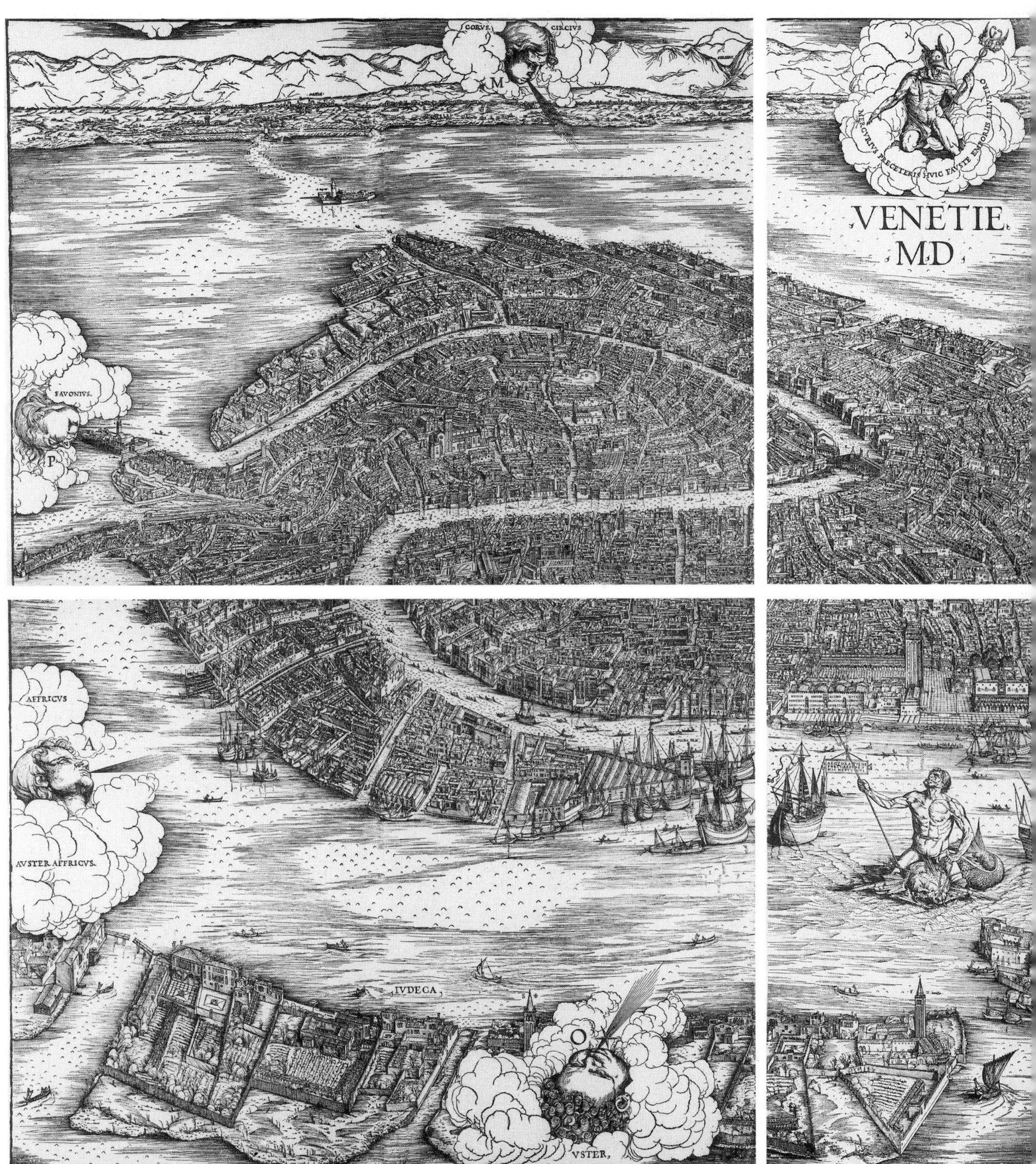

military surveyors produced the first scale plans of Italian towns to be made since Roman times. Leonardo da Vinci's plan of Imola, *c.*1502, pioneered the use of scaled ground plans for military purposes, which became common throughout northern Europe during the first half of the sixteenth century. Military considerations also determined the work of the surveyor Jacob van Deventer, who obtained a commission from Philip II of Spain to produce ground plans to a uniform scale of the walled

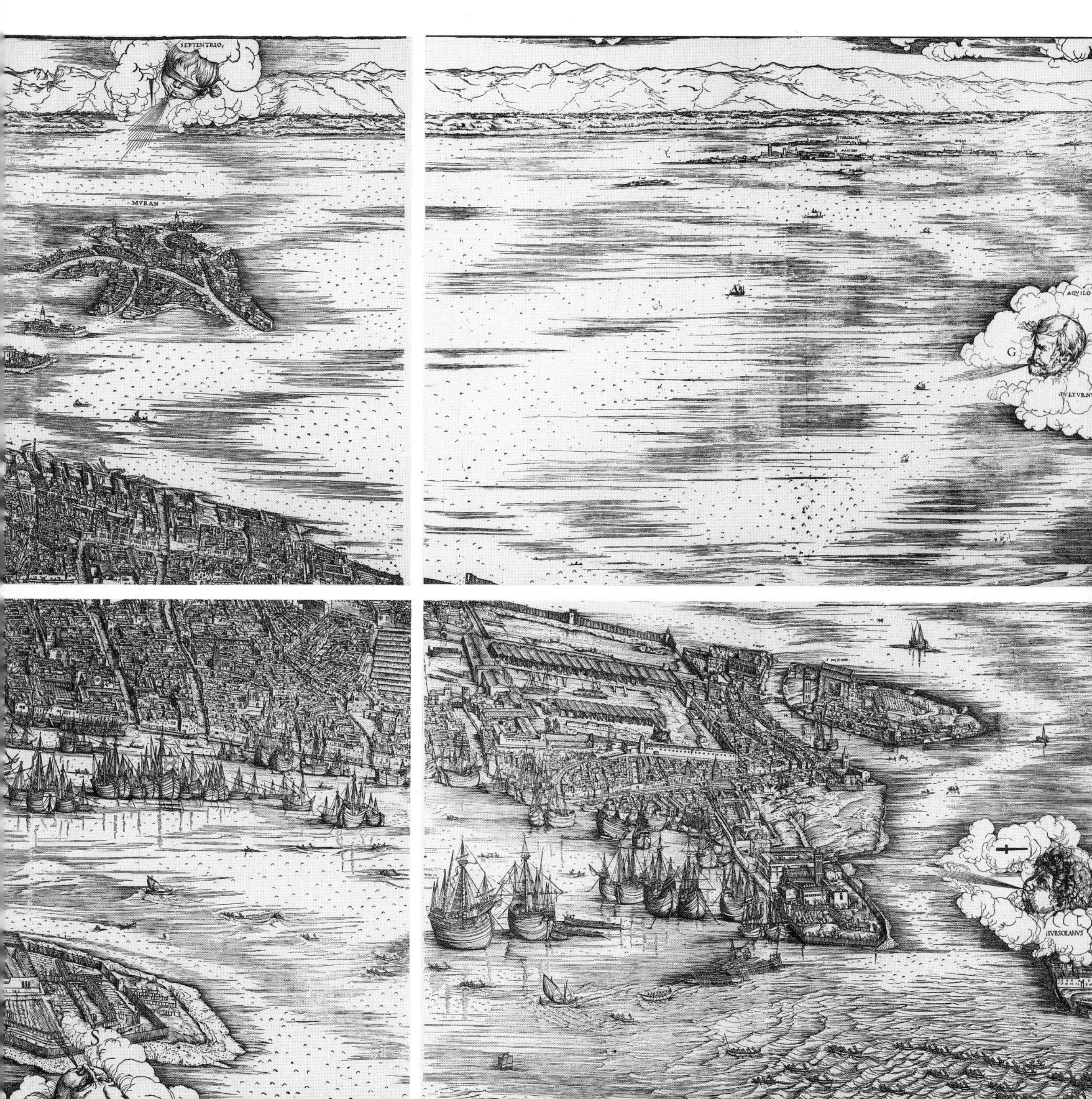

(11) Jacopo de' Barbari, woodcut view of Venice. (Venice, 1500.) British Museum: Department of Prints and Drawings, 1895-1-22-1192. Venice, then at the height of its power, is shown from the south-west. The famous Campanile in the Piazza S Marco is visible (centre) with a temporary flat roof erected after the fire of 1493. The view was subsequently revised to show the conical roof restored in 1513.

towns of the Netherlands, some 250 altogether. His work, which began in 1558, continued until 1572, and although his plans were official documents not at first intended for publication, a number of them were engraved on copper to contribute to the variety of cartographic styles found in the pages of Braun & Hogenberg's *Civitates Orbis Terrarum* of 1572–1617.

(12) 'Le sette chiese di Roma' (Rome: Antonio Lafreri, 1575.) BL Maps 23807.(1).

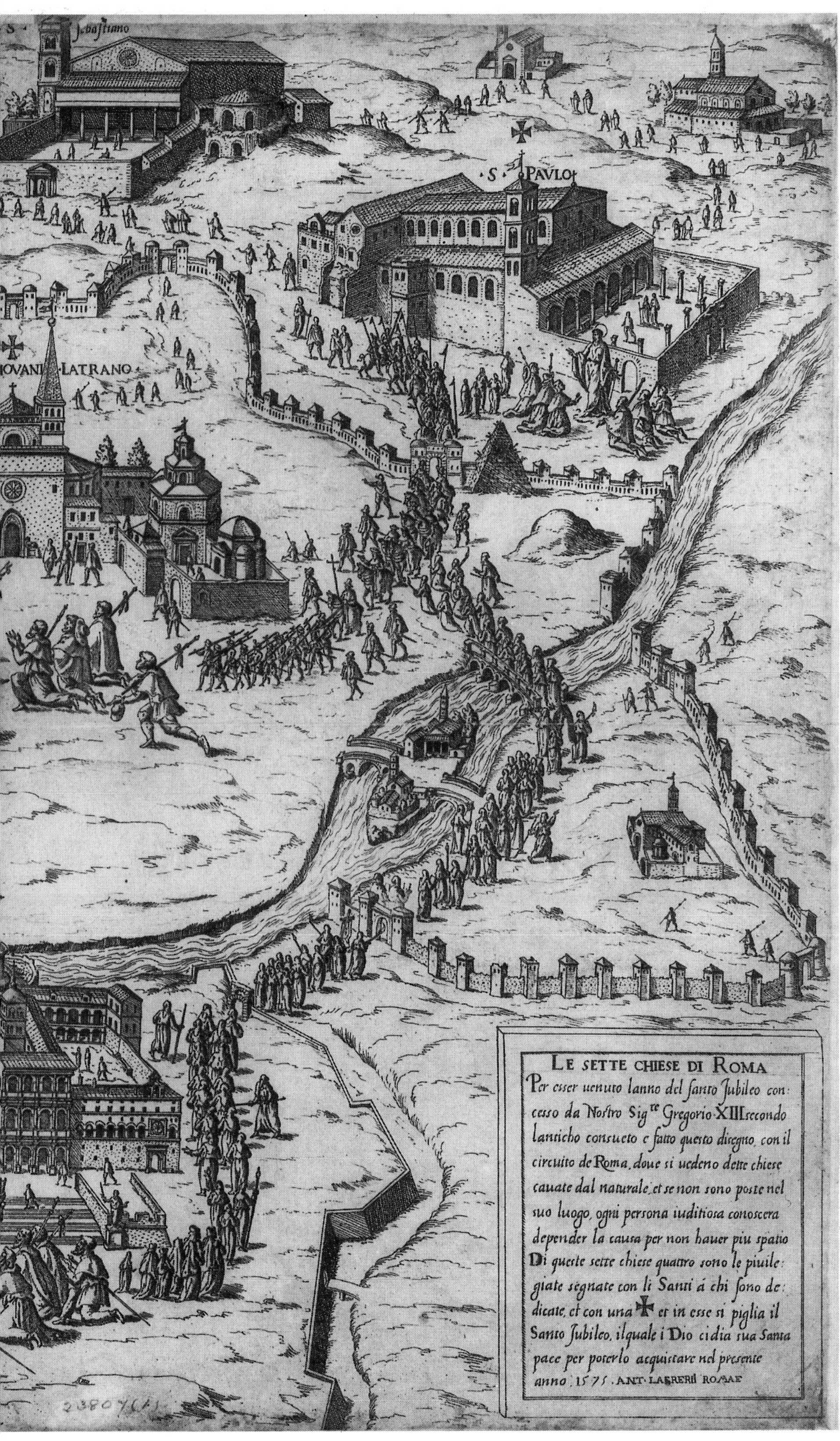
S. bastiano
S. PAVLO
OVANI LATRANO
LE SETTE CHIESE DI ROMA
Per esser uenuto lanno del santo Jubileo con:
cesso da Nostro Sig.re Gregorio XIII secondo
lanticho consueto e fatto questo disegno, con il
circuito de Roma, doue si uedeno dette chiese
cauate dal naturale, et se non sono poste nel
suo luogo, ogni persona iuditiosa conoscera
depender la causa per non hauer piu spatio
Di queste sette chiese quattro sono le piuile:
giate segnate con li Santi á chi sono de:
dicate, et con una ✠ et in esse si piglia il
Santo Jubileo, ilquale i Dio ci dia sua Santa
pace per poterlo acquistare nel presente
anno. 1575. ANT. LAFRERII ROMAE

2

Braun & Hogenberg and after: the town plans of the seventeenth century

Until the late sixteenth century, the portrayal of cities was best known through paintings and illuminated manuscripts, book illustrations, broadsheets, large-scale bird's-eye views or details on topographical maps. In 1572, however, the first volume of the *Civitates Orbis Terrarum* was published in Cologne. The sixth and final volume appeared in 1617. This great city atlas, edited by Georg Braun and largely engraved by Frans Hogenberg, eventually contained 546 prospects, bird's-eye views and map-views of cities from all over the world. Braun (1541–1622), a cleric of Cologne, was the principal editor of the work, and was greatly assisted in his project by the close and continuing interest of Abraham Ortelius, whose *Theatrum Orbis Terrarum* of 1570 was, as a systematic and comprehensive collection of maps of uniform style, the first true atlas. The *Civitates*, indeed, was intended as a companion for the *Theatrum*, as indicated by the similarity in the titles and by contemporary references regarding the complementary nature of the two works. Nevertheless, the *Civitates* was designed to be more popular in approach, no doubt because the novelty of a collection of city plans and views represented a more hazardous commercial undertaking than a world atlas, for which there had been a number of successful precedents. In a letter to Ortelius dated 31 October 1571, Braun explains that:

(13) 'Palma'. From Georg Braun and Frans Hogenberg, *Civitates Orbis Terrarum*, book V. (Cologne, 1598 [edition of 1623?])
BL Maps C.29.e.1.
Palmanova was one of the purest examples of a planned Renaissance town. The radial-concentric design emphasised by the plan-view published in the *Civitates* was primarily for military purposes. The use of artillery in siege warfare had rendered stout curtain-walls obsolete, and the defences of Palmanova depended instead on a series of strong points backed up by covering fire from a central point.

> For various reasons some learned men here in Cologne think that Master Frans's Book of Cities would commend itself more to purchasers if the proper names of places, churches and gates were given in the native language, so as to satisfy both the learned and the unlettered: the learned because they will have the Latin descriptions on the back, the unlettered because each will see his own native town depicted with places named in a form familiar to him. I think the usefulness of this arrangement evident when pictures of cities are sold separately, but the citizens would like them less if they could understand nothing of what they read. For while this is nevertheless the practice in the descriptions [i.e. maps] of countries in your very beautiful Theatrum, and admitting its usefulness to the learned, yet you see how many people are distracted by it; and since Master Frans's subject is a popular one, delighting the eyes even of the unlettered, it would appear sensible to me to accommodate ourselves somewhat to them . . . these are matters, Master Abraham, which I thought of most importance to you and on which no decision will be taken until we know your considered opinion.

'Master Frans', Frans Hogenberg, was the son of a Munich engraver who settled in Malines. He engraved most of the plates for Ortelius's *Theatrum* and the majority of those in the *Civitates*, and may have been responsible for originating the project, as the phrase in Braun's letter 'Master Frans's book of Citties' would seem to imply.

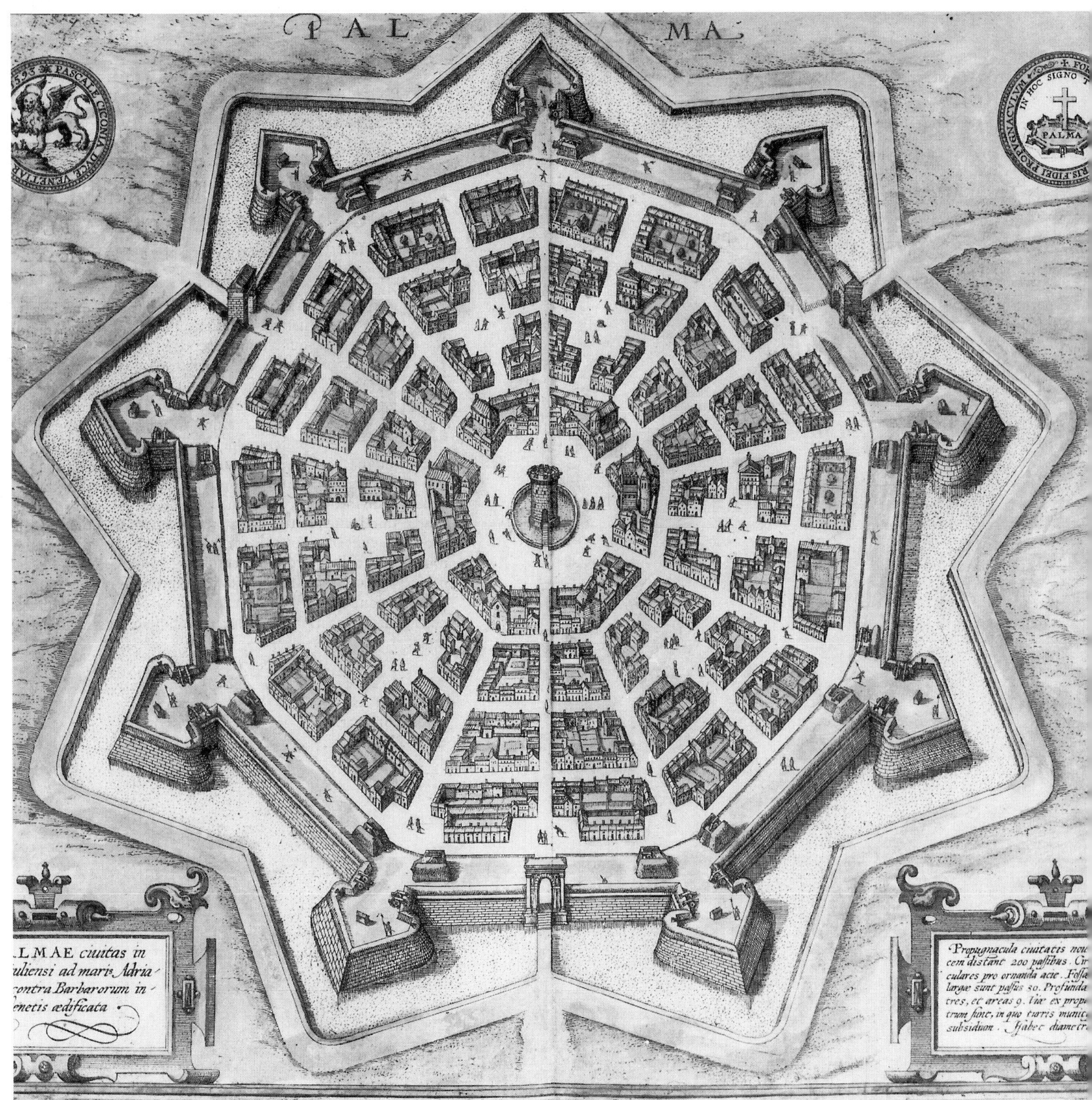

The copper-plates of the *Civitates* were engraved from drawings by a number of different artists and cartographers, the most significant of whom was Georg Hoefnagel. He not only contributed most of the original material for the Spanish and Italian towns but also reworked and modified those of other contributors. A large number of Deventer's plans of towns of the Netherlands were copied, as were Stumpf's woodcuts from the *Schweizer Chronik* of 1548, and Münster's German views from the 1550 and 1572 editions of his *Cosmographia*. The *Civitates* provided a uniquely comprehensive view of urban life at the turn of the sixteenth

(14) Hans Lautensack, 'Warhafftige Contrafactur der löblichen Reychstat Nuremberg, gegen dem Nidergang der Sonnen'. (Nuremberg, 1552.) BL Maps 28840.(7).
Hans Lautensack produced prospects of Nuremberg, from the east and from the west, which are noted for the reliability of their topographic detail. Both feature a self-portrait in which the artist, surrounded by curious bystanders, is identified by his monogram.

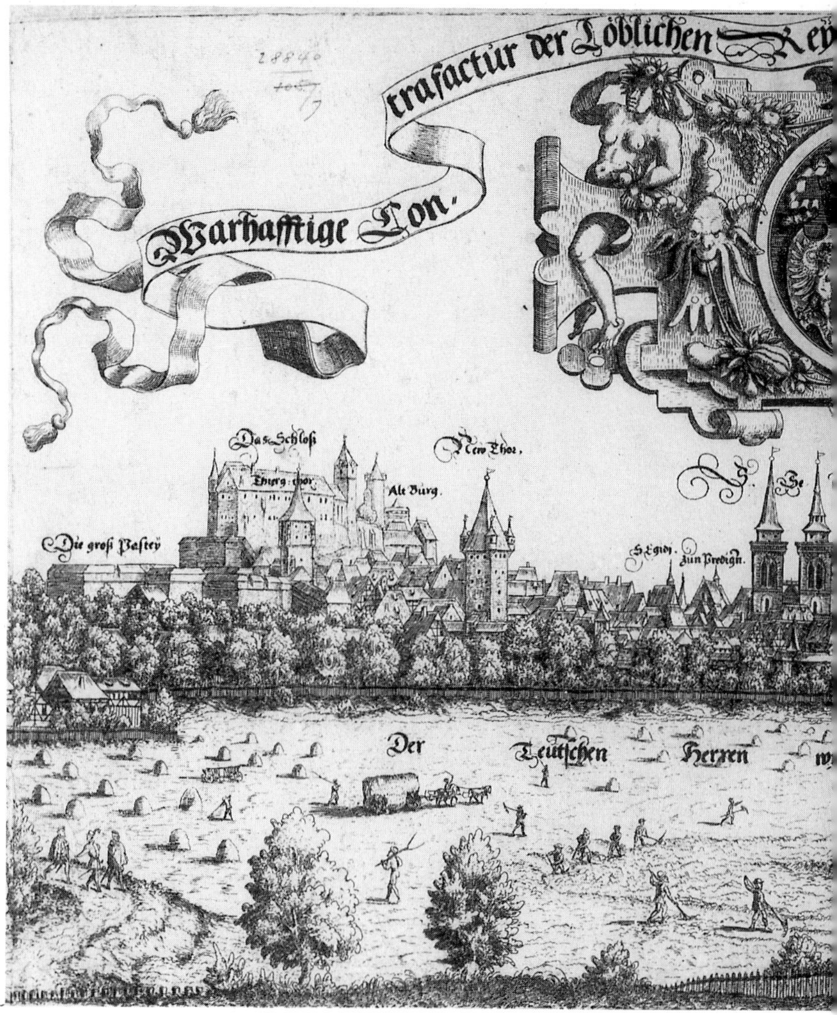

century. Braun's stipulation that 'towns should be drawn in such a manner that the viewer can look into all the roads and streets and see also all the buildings and open spaces' can be demonstrated in the map-view of Cologne in book I (1572) (PLATE 3) and the plan of Palmanova from book V (13). In common with the other plans and views in the *Civitates*, both are decorated with figures in local dress. This feature was anticipated in Hans Lautensack's etched view of Nuremberg, 1552 (14), whose groups of citizens in the rural foreground add further authenticity to the highly accurate topographical details of what was effectively Germany's cultural capital at that time. Braun's motives for adding figures to the views, however, went further: as stated in his introduction to book I, he believed, perhaps optimistically, that his plans would not in consequence be scrutinised for military secrets by the Turks, as their religion forbade them from looking on representations of the human form. The plans, each accompanied by Braun's printed account of the town's history, situation and commerce, form an armchair traveller's compendium, which the scholar Robert Burton in *The Anatomy of Melancholy* of 1621 asserted would not only provide instruction but would uplift the spirit as well.

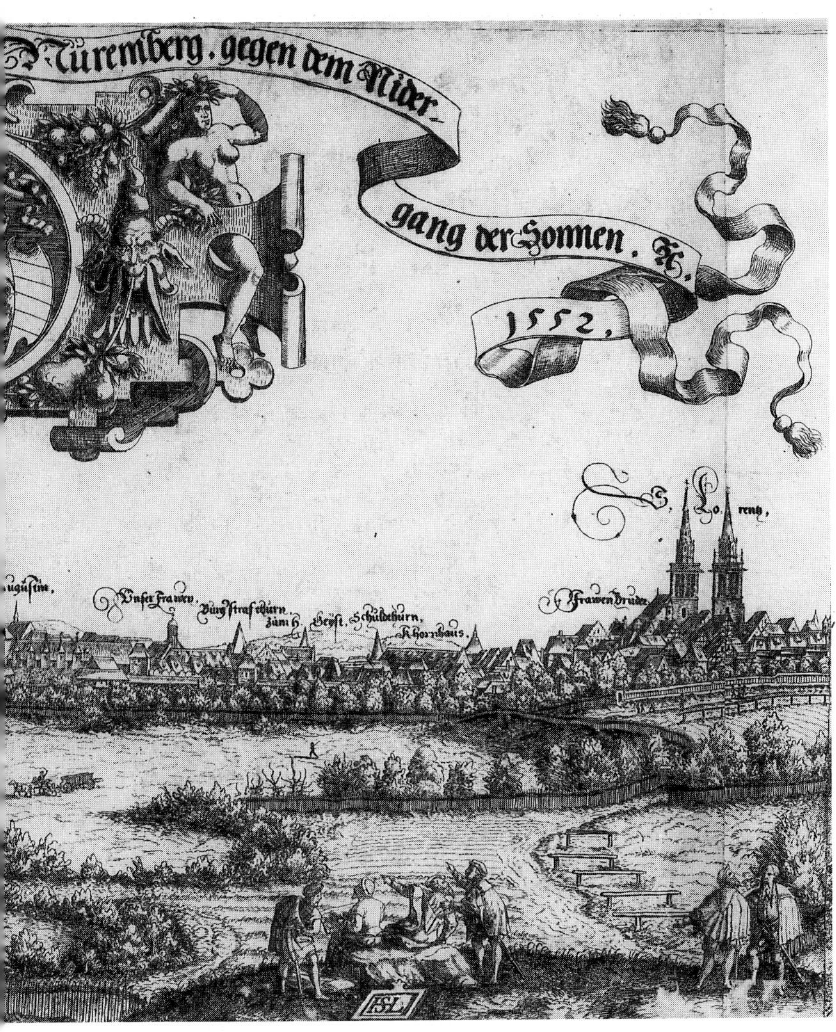

The success of the *Civitates* started a fashion for 'townbooks' that was to last for the rest of the seventeenth century and into the eighteenth century. Braun and Hogenberg invited other authors to copy their work, an offer taken up by Matthäus Merian the elder, who reproduced Braun and Hogenberg's views sixty years later in his *Theatrum Europaeum*, a 21-volume work of topographic views published from 1640 onwards. The Amsterdam map publisher Jan Jansson purchased 363 of the plates of the *Civitates* in 1653, using 232 of them with little change other than to the imprint for his eight volumes of town views issued from 1657 onwards (**15**). This work contained 500 plates, almost all of European towns, and incorporated, in addition to those from the *Civitates*, 173 newly-engraved plates reflecting the more austere style of the mid seventeenth century. Jansson derived a number of these from Joan Blaeu's two-volume townbook of the Netherlands, published in Amsterdam in 1649 (**16**).

The history of Blaeu's townbooks is intimately linked with the Dutch war of independence from Spain. Blaeu originally intended that both volumes should contain 110 plates, Volume I for the towns and fortresses in the Seven United Provinces, then struggling for independence, and

(15) 'Argentina. Strassburg'. From Jan Jansson, *Theatrum exhibens illustriores principesque Germaniae Superioris civitates*, Volume I. (Amsterdam, 1657.)
BL Maps C.25.b.15. [Detail.]
The view of Strasbourg was one of a minority of plates for which Jansson commissioned a new engraving for his townbooks, the majority being made up of plates acquired from the *Civitates* of Braun and Hogenberg. Jansson's own plates reflect the noticeably more austere style of the mid-seventeenth century.

(16) 'Legia sive Leodium, vulgo, Liege'. From Joan Blaeu, *Toonneel der Steden van's Konings Nederlanden*. (Amsterdam, 1649.)
BL Maps C.4.d.3.

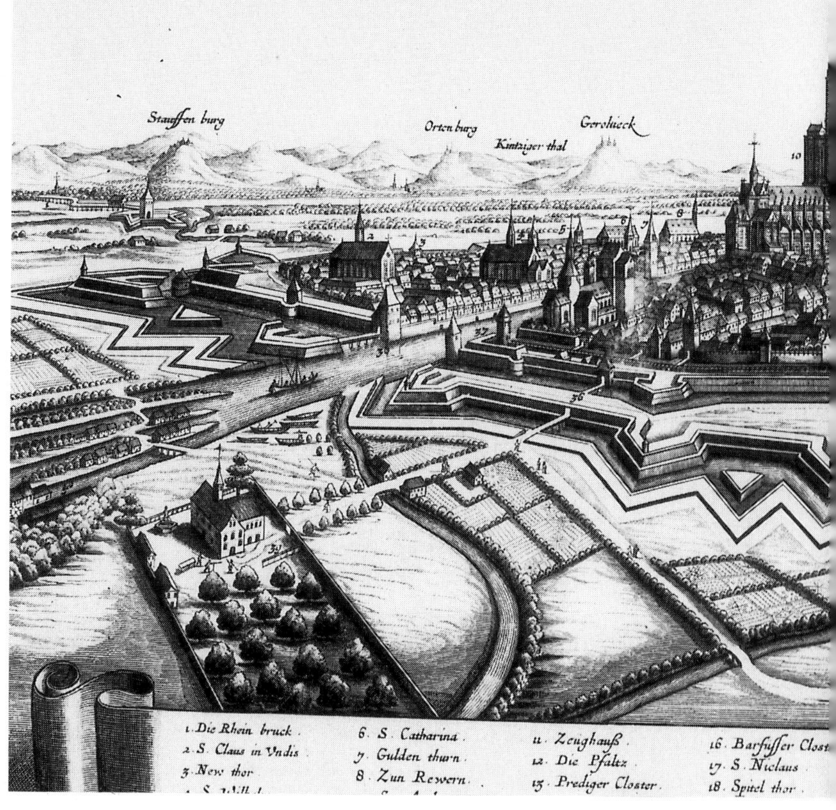

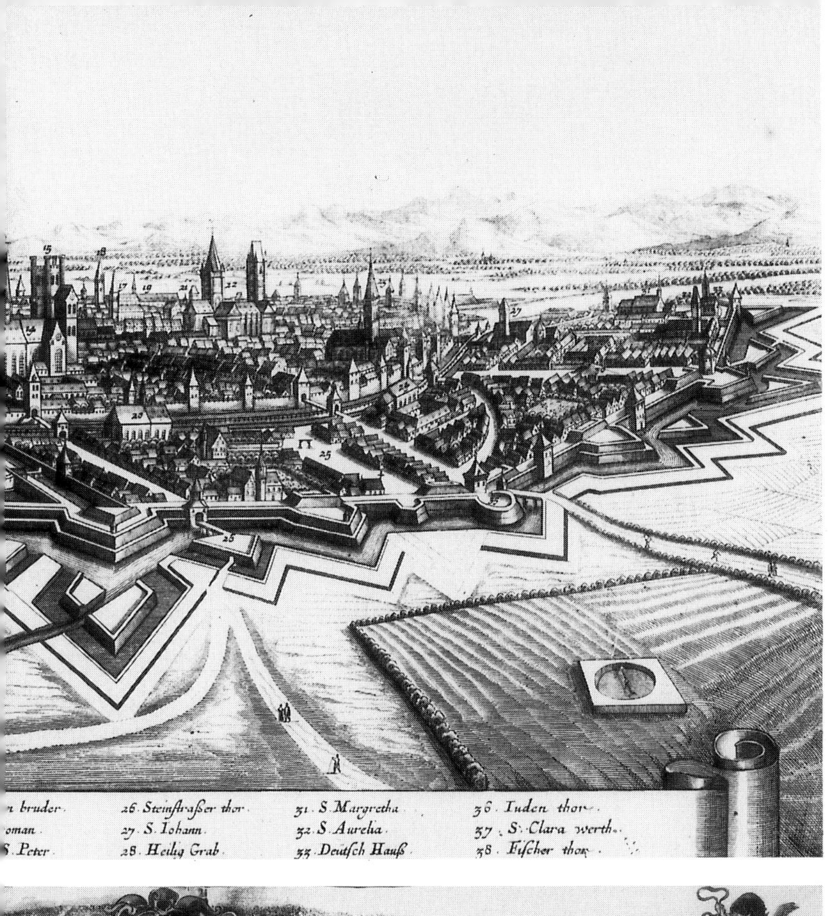
n bruder. 26. Steinstraßer thor. 31. S. Margretha. 36. Iuden thor.
oman. 27. S. Iohann. 32. S. Aurelia. 37. S. Clara werth.
S. Peter. 28. Heilig Grab. 33. Deutsch Hauß. 38. Fischer thor.

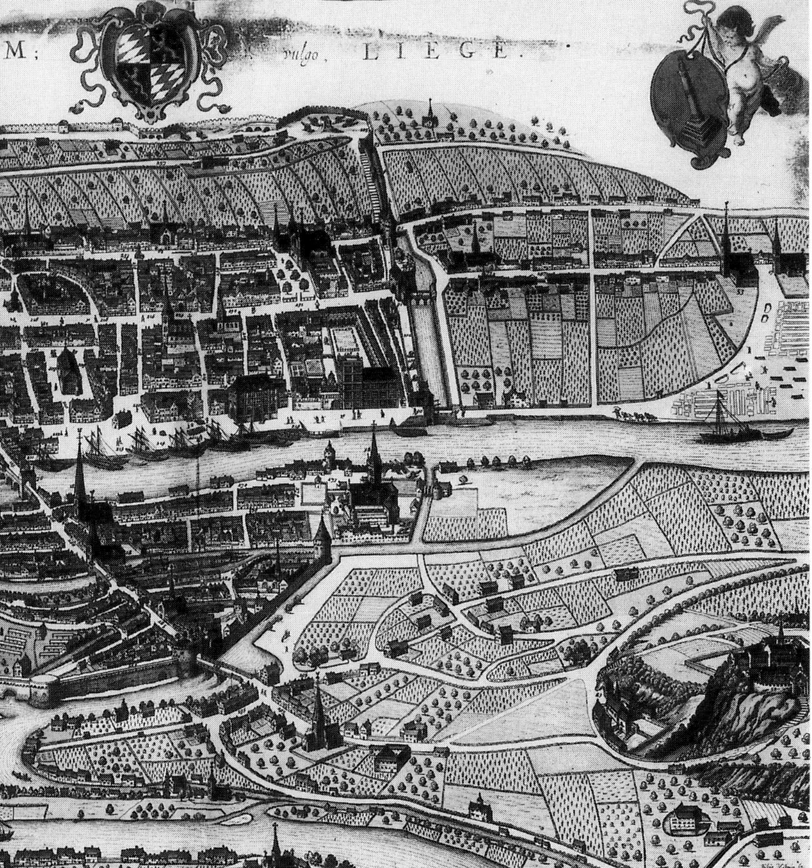
M; vulgo, LIEGE.

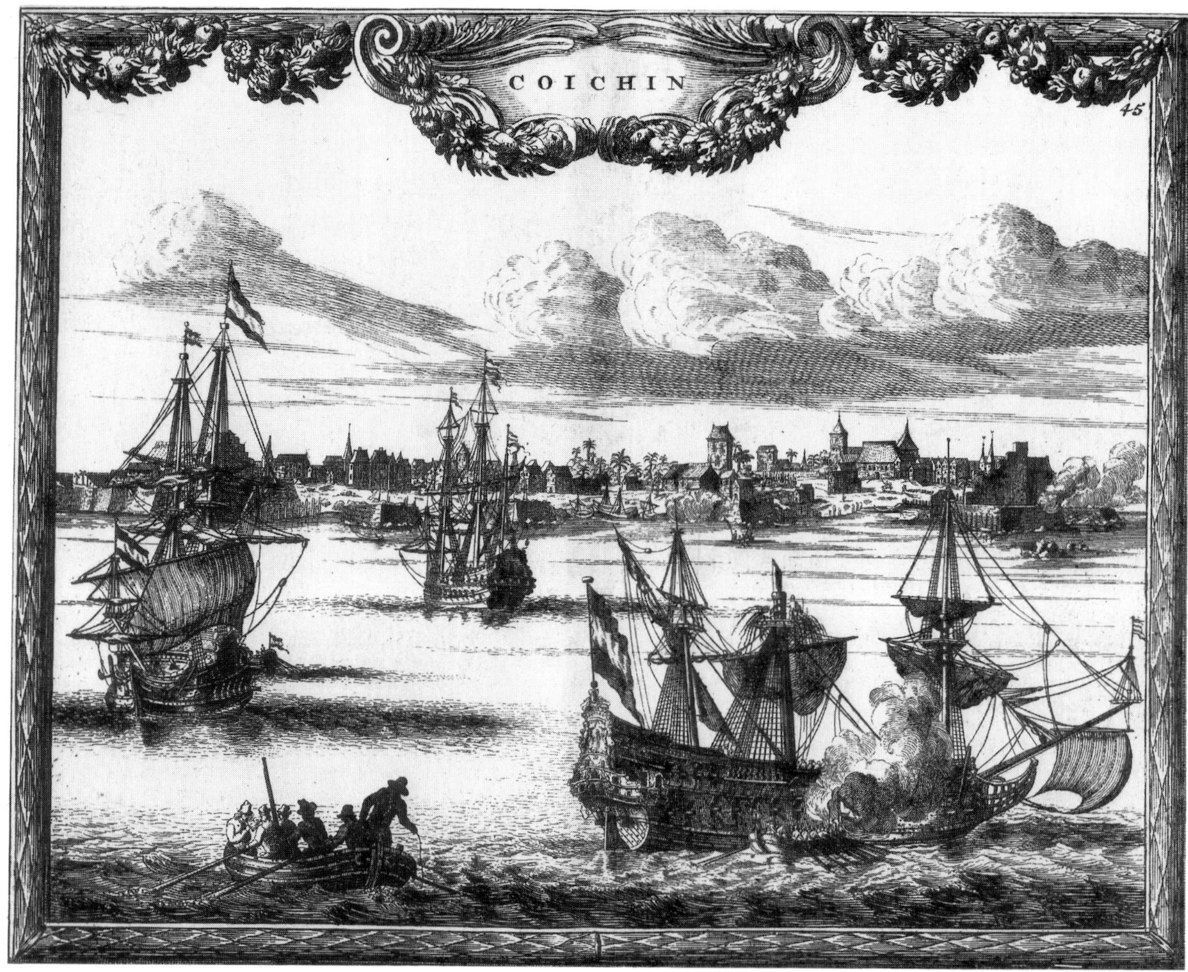

(17, 18) Paired views of Cochin and its inhabitants. From Carel Allard, *Orbis habitabilis oppida et vestitus*. (Amsterdam, *c*.1695.) BL Maps C.24.e.23.

Volume II for the towns controlled by Spain. By the time the first edition of the atlas was ready for publication, a number of towns had changed hands and their new allegiances had been settled by the Treaty of Westphalia of 1648. For the second and subsequent editions, Blaeu rectified the confusion by moving the plates of the 26 towns captured by the United Provinces to the first volume, and by transferring to the second volume plates of five towns captured by Spain.

Other collections of town plans and views, similar in style to the *Civitates* and in many cases derived directly from them, continued to appear into the early eighteenth century. A relatively late example, which elaborated upon the theme of Braun and Hogenberg, was the *Orbis habitabilis oppida et vestitus* by Carel Allard (17, 18). His collection (published in Amsterdam in 1695) consisted of a number of paired plates, one of which was a prospect, the other of which prominently featured a couple dressed in local costume with a prospect serving as a background. In this way, the reader could 'travel the world . . . discoursing with remote peoples, admiring their costumes and ornaments, their churches and buildings'.

Although the townbooks gave European scholars their first glimpses of the non-European city, the printing of town plans was by no means

confined to Europe. Amongst the first Japanese printed materials to reach the west was an anonymous map of Nagasaki (19) brought back to Europe by Englebert Kaempfer (1651–1716), a German who in 1690 became physician to the embassy sent by the Dutch East India Company to the Shogun's court in Edo (Tokyo). By this time the Togukawa Shogunate had banished Christianity, and only the Chinese and Dutch were permitted to trade with Japan, through the port of Nagasaki. Here, the Dutch maintained a small trading settlement on the island of Deshima. The map of the city, executed in *c.*1680, was typical of Japanese *toshizu* (city maps) in that it was designed to be read on the floor and thus meant to be viewed from four different directions.

The popularity of the printed townbook did not cause the demise of manuscript town plans. These continued to fulfil an important role, especially for military purposes, affairs of state and for presentation. The 'Livro do Estade da India Oriental' prepared by Antonio Boccaro in 1646 (PLATE 4) was an expanded version of a survey of Portugal's African and Asian colonial empire which had originally been commissioned in 1632 by Philip III of Portugal (better known as Philip IV of Spain). It was copiously illustrated with plans and views of the major towns, forts and trading posts of the empire by Barreto de Resende, who was accused by

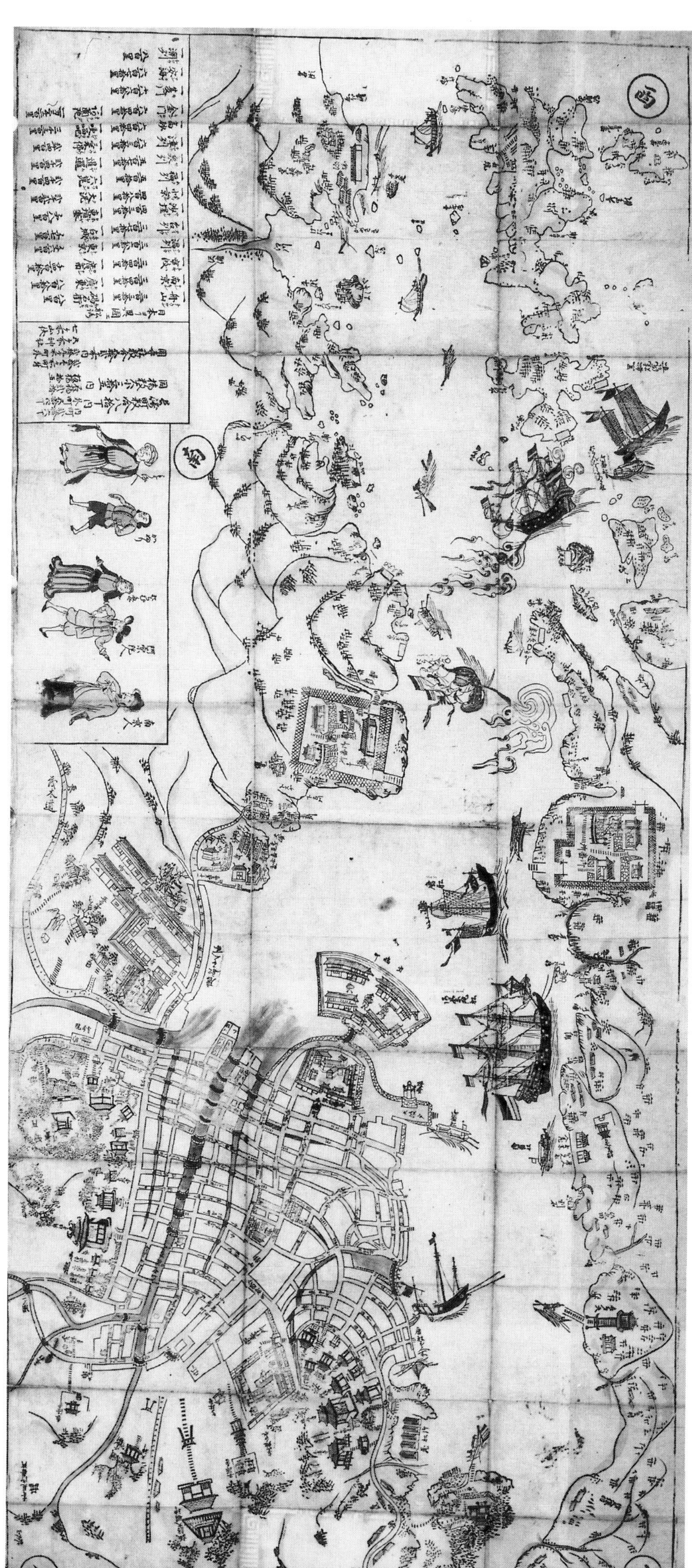

(19) Anonymous woodcut map of Nagasaki, *c.*1680.
BL OMPB Oriental 75.g.25.
The small Dutch trading settlement on the artificial island of Deshima can be seen at centre, also the Chinese station on the outskirts of the city, at left. The harbour, containing several Dutch and Chinese vessels among its shipping, is guarded on either side of its mouth by forts containing detachments of the Japanese Imperial Guard. The map is decorated with figures of foreigners, and contains a table of distances to several other countries, including Holland.

(PLATE 3) Plan of Cologne. From Georg Braun and Frans Hogenberg, *Civitates Orbis Terrarum*, book I. (Antwerp and Cologne, 1572 [edition of 1623?].)
BL Maps C.29.e.1.
Cologne, the home town of Braun and Hogenberg and the birthplace of the *Civitates*, was depicted by means of a map-view, in which the major buildings were drawn in elevation on an ordinary ground plan. The cathedral, destined to remain unfinished for a further three centuries, and the city walls, since demolished to make way for the Ringstrasse, are clearly visible.

(PLATE 4) Plan of Goa by Pedro Barreto de Resende. From Antonio Boccaro, 'Livro do Estade da India Oriental' (Book of the State of East India), 1646.
BL Sloane MS 197, ff.247v–248.
Goa was the capital of Portugal's eastern empire. Captured in 1510, it was not finally abandoned by the Portuguese until 1961. It was an elegant city, of which it was said that 'he who has seen Goa need not see Lisbon'.

The viceregal palace where Resende worked, can be seen at the centre behind the tree on the square.

ORIENS
COLONIA AGRIPPINA
OCCIDENS

GOA
CABO
BARDES

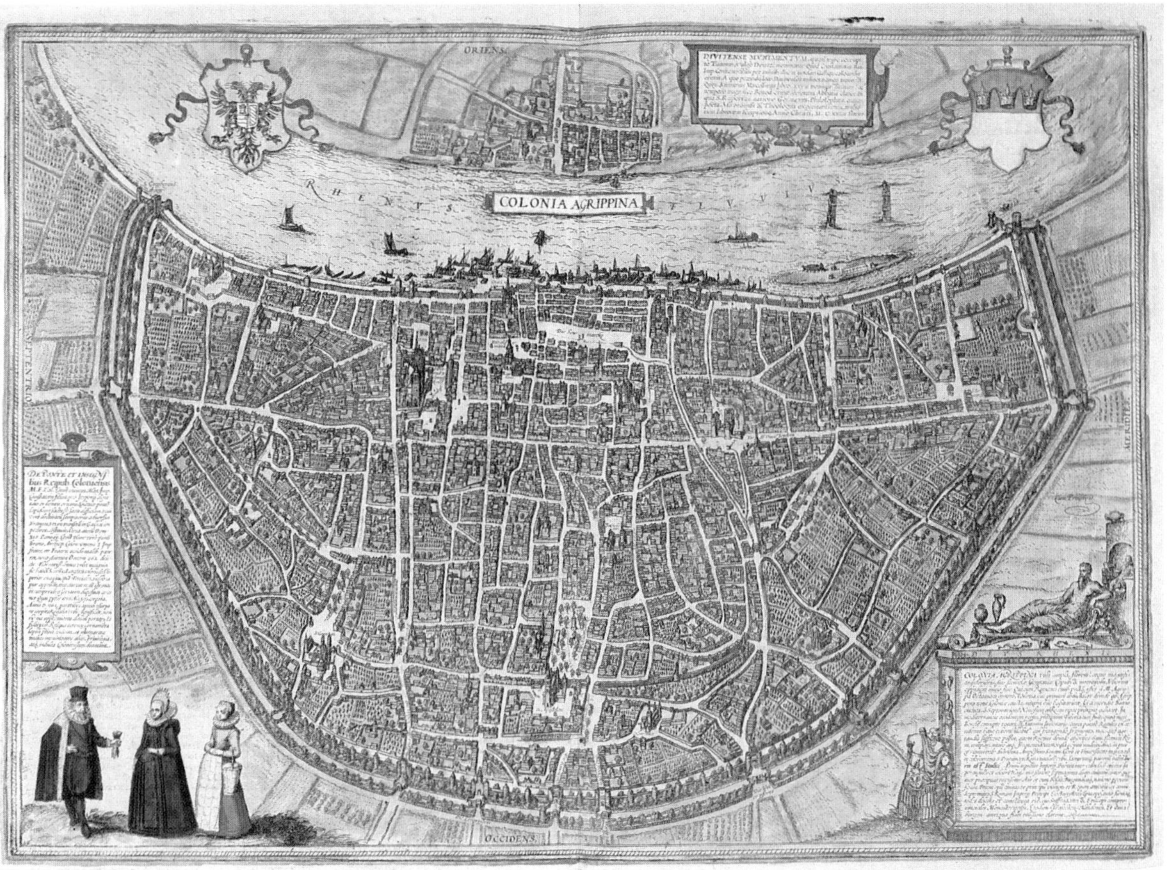

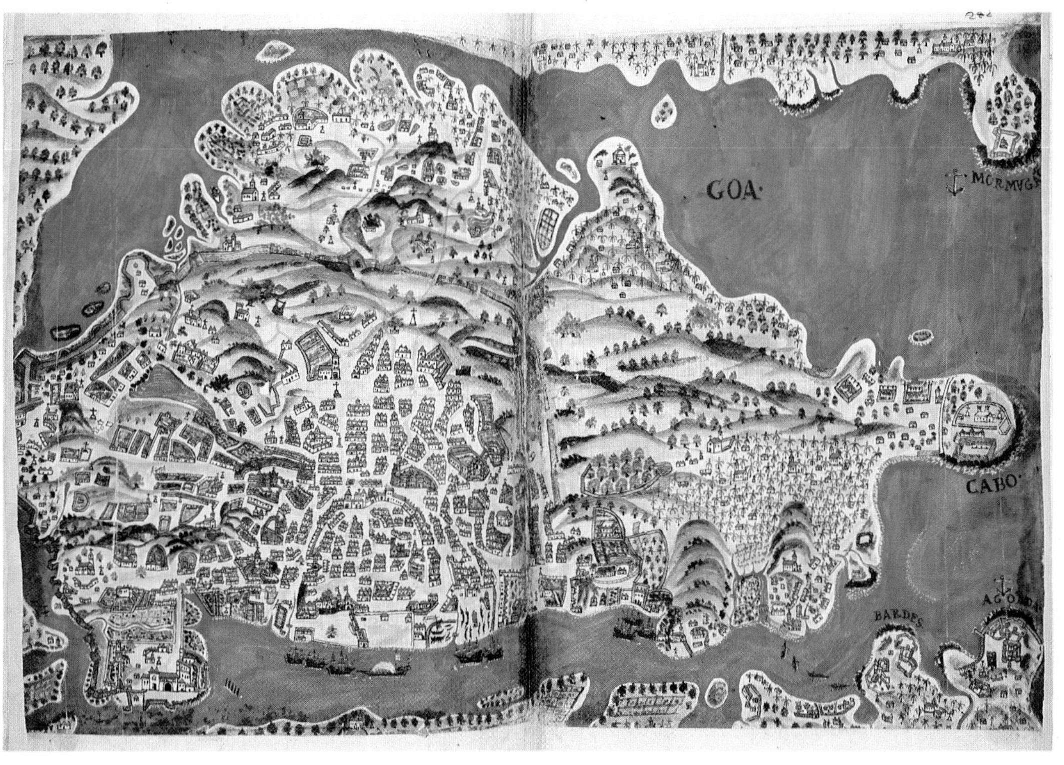

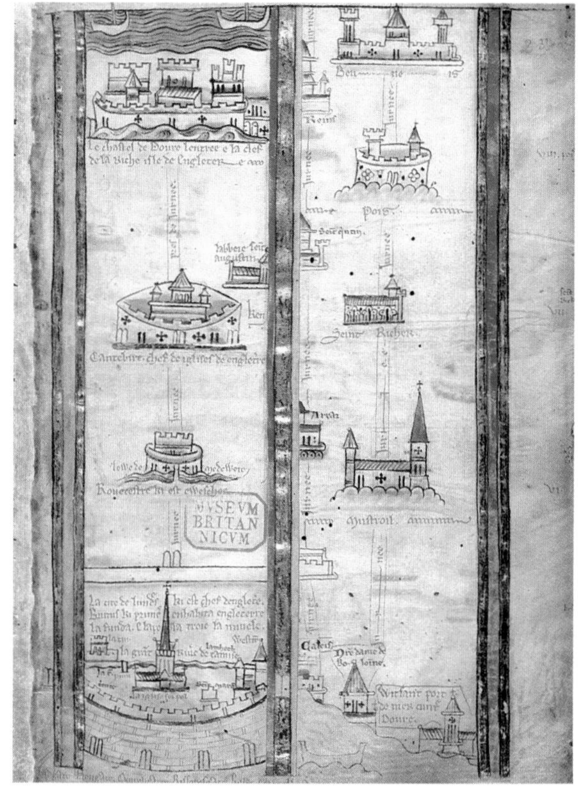

(PLATE 5) Introductory section of the itinerary from London to Apulia. From Matthew Paris, 'Historia Anglorum', made after 1252.
BL Royal MS 14.C.VII, f.2.
The itinerary, containing one of the earliest known depictions of London, shows the city as viewed from the north (from Hampstead Heath?). Clearly visible are old St Paul's Cathedral ('la iglise sainct pol'), Westminster Abbey (incorrectly placed within the city walls), the Tower ('la tur') and the River Thames ('la grant Riviere de tamise') crossed by London Bridge ('punt').

Boccaro of executing the plans without proper scientific measurement. Resende, who chose the time-honoured method of the bird's-eye view and map-view, was more concerned with depicting the town of Goa in an artistic style. He seems also to have been influenced by native Indian cartography, as the use of thick, vivid colour and the crowding together of groups of stylised minor buildings may testify. Perhaps Boccaro found this style somewhat old-fashioned and oriental.

A British example, possibly prepared for James, Duke of York, and traditionally known as the 'Duke's plan', shows the city on the southern tip of Manhattan Island then known as New Amsterdam (see cover picture). Originally a fort established by the Dutch in 1625, the settlement had developed into a flourishing port and trading city by 1664, when it was captured by the British and renamed New York. The title of the plan refers to September 1661, and suggests that the cartographer who executed the anonymous English plan three years later had access to the second of two surveys made in 1660 and 1661 by the Dutchman Jacques Cortelyou. The plan of 1661 may have been handed over to the English along with the town by the Governor, Petrus Stuyvesant.

Tangier, another recently acquired colony, was depicted by John Seller, who included a plan of the town in his *Atlas Maritimus, or the Sea Atlas*, published in 1675. It had been a British possession from 1662, when Catherine of Braganza gave it to Charles II as part of her dowry, until 1683, when it was evacuated.

(PLATE 6) John Hooker, 'Isca Damnoniorum' (Exeter). Engraved by Remegius Hogenberg. (London, 1587.) BL Maps C.5.a.3.
Hooker's bird's-eye view of Exeter reveals the compact nature of the Tudor city as well as its close connections with the surrounding countryside. The 8000 inhabitants lived in an area of only 93 acres, while the presence of fields and burgage plots (allotments) demonstrate the 'rus in urbe' nature of the pre-industrial town. The River Exe is portrayed as an important source of livelihood, with fishing, milling and shipping represented.

3

Early plans in the British Isles

As on the European continent, the first representations of towns and cities in the British Isles were stylised pictorial drawings, some of the earliest of which can be found in the itinerary (PLATE 5) executed some time after 1252 by Matthew Paris (*c.*1200–1259), a monk at St Albans' Abbey. The itinerary was intended to illustrate the journey made to southern Italy by Henry III's brother, Richard, Earl of Cornwall, on his election in 1252 as King of Sicily, and appears as part of the 'Historia Anglorum', an abridgement of Paris's earlier 'Chronica Maiora'. The first section of the itinerary, depicting the route from London to Beauvais, shows the road as a straight line, with the towns of London, Rochester, Canterbury and Dover placed in the order in which they were encountered by the traveller rather than in their correct spatial relationship. In this respect, Paris's map anticipated the work of John Ogilby, whose *Britannia* of 1675 repeated the strip-map technique. The representational method employed by Paris can be detected today in the route maps produced by motoring organisations. The towns on the itinerary were depicted in perspective and, in the case of London and Canterbury, were shown surrounded by their walls in a manner that harks back to the classical traditions of some versions of the *Corpus Agrimensorum*. In the case of London, the walls were generalised, but the identity of the city was established by drawing and naming prominent buildings, such as St Paul's Cathedral and the Tower.

The earliest known English plan believed to result from the application of scientific methods of survey to an urban ground plan appears in a manuscript portrayal of Bristol, then England's second largest city, on folio 5 of 'The Maire of Bristowe is Kalendar', compiled by Robert Ricard in *c.*1479. Ricard, then the city's town clerk, depicted Bristol as a bird's-eye view with the High Cross at the centre of four wide streets, each of which seems to have been drawn along broadly correct alignments to one of the city's four gates.

The application of printing techniques to urban cartography did not occur until the middle of the sixteenth century, when plans and views of the major urban centres became available in small but steadily increasing numbers. London, whose population of about 150,000 in 1559 amounted to some 10 per cent of England's total, was the most common subject. The earliest known plan of the city was one originally executed on fifteen copper-plates by an unknown artist at a scale of about 34 inches to 1 mile (*c.*1:1860). The two surviving plates, both now in the Museum of London, are datable from internal evidence to between 1553 and 1559, and cover a north-south section of the city from Moorgate to the Thames. No known impression of the so-called 'copper-plate map' now survives, but it was evidently copied by the author of a woodcut map on eight sheets once attributed to the Elizabethan surveyor Ralph Agas (1545–1621). The three surviving copies of the 'Agas map' seem to have been produced some time between 1561, when the spire of St Paul's Cathedral was destroyed, and 1570, when the new Royal Exchange was built. The 'copper-plate map' was also copied by Frans Hogenberg for his map-view of the city for book I of the *Civitates Orbis Terrarum* of 1572.

Almost contemporary with the 'copper-plate map' is a woodcut bird's-eye view of Norwich dated 1558, which thus forms the earliest surviving printed map of known date of any English town (**20**). The author, William Cuningham, was a distinguished physician and astrologer who included the map as one of several woodcuts illustrating principles of map-construction in *The Cosmographical Glasse* of 1559, one of the first treatises on practical map-making published in England. Cuningham's view was subsequently redrawn by William Smith for book III of the *Civitates* of Braun and Hogenberg, 1581, and for Smith's own manuscript account of England of 1588. It appears also to have been the basis for John Speed's depiction in his *Theatre of the Empire of Great Britaine*, 1611–12. Norwich was then the second or third largest city in England, and a short verse, concluding the description in the *Civitates*, reads:

Norwich may rightly pay respects to London's prior claim,
and offer reverence to York of venerable fame,
But doubt if justice bids her yield to Bristol all the same.

The earliest depiction of a town known to be engraved on copper by an Englishman was the bird's-eye view of Cambridge by Richard Lyne. This is commonly found with the second edition of *Historia Cantebrigiensis Academiae*, 1574, which was published by John Caius in the course of the simmering controversy as to whether Oxford or Cambridge was the older university. The book was requested by Matthew Parker, Archbishop of Canterbury, himself a former Cambridge scholar. Parker employed both Lyne and Remegius Hogenberg, whose brother Frans co-operated with Georg Braun in the publication of the *Civitates*. Remegius also engraved some of the plates for Christopher Saxton's *Atlas of England and Wales* (1579), as well as another bird's-eye view of an English provincial city, that of Exeter in 1587 (PLATE 6). John Hooker, the surveyor and author of the view, was Chamberlain to the City of Exeter. In this capacity, he was responsible for the safety of the city's official records, arranging for the care of the orphans, and surveying the city's property. He was also an antiquary and historian of national stature, who revised and edited the second edition of Holinshed's *Chronicles*, in addition to publishing numerous works relating to his native Exeter. His survey was probably inspired by both his works as an historian and his status as an official. The view, the latest on which Remegius Hogenberg's name appears, was later copied by Speed for his *Theatre*.

The Cheshire-born herald and topographer William Smith (*c*.1550–1618) contributed much to the urban mapping of England with his manuscript 'The Particuler Description of England' of 1588 (**21**). This unfinished work, apparently designed for publication, was arranged

(**20**) Woodcut map of Norwich. From William Cuningham, *The Cosmographical Glasse*. (London, 1559.)
BL 59.i.28.
The city is viewed from the west, with the surveyor (Cuningham) and his plane-table in the foreground. Places of interest are lettered and referred to by a table on the reverse of the map. Location 'C', outside the walls to the top of the map, was 'the place where men are customablie burnt', a reference to the severe persecution of Protestant heretics in Norwich between 1554 and 1559.

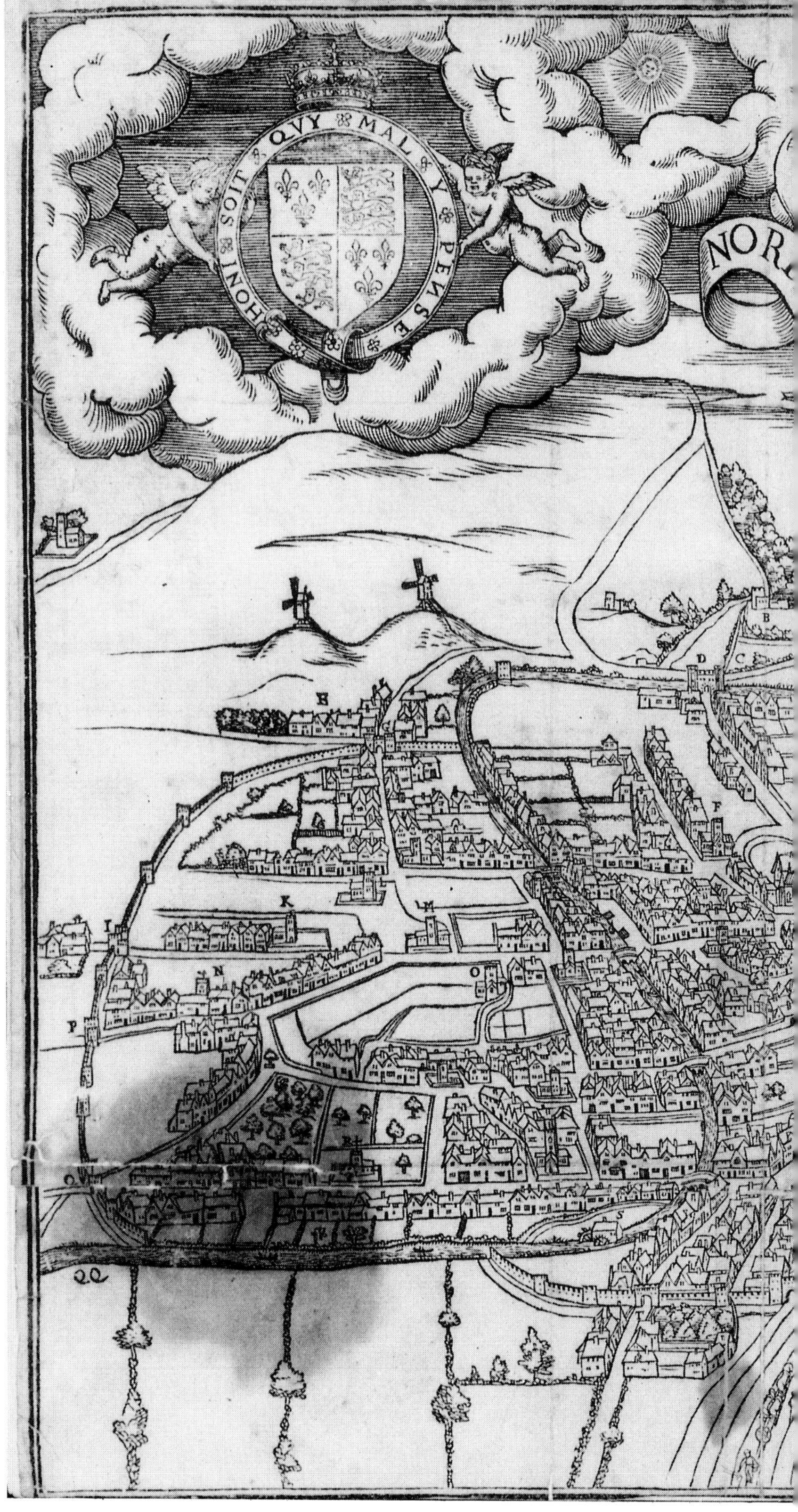

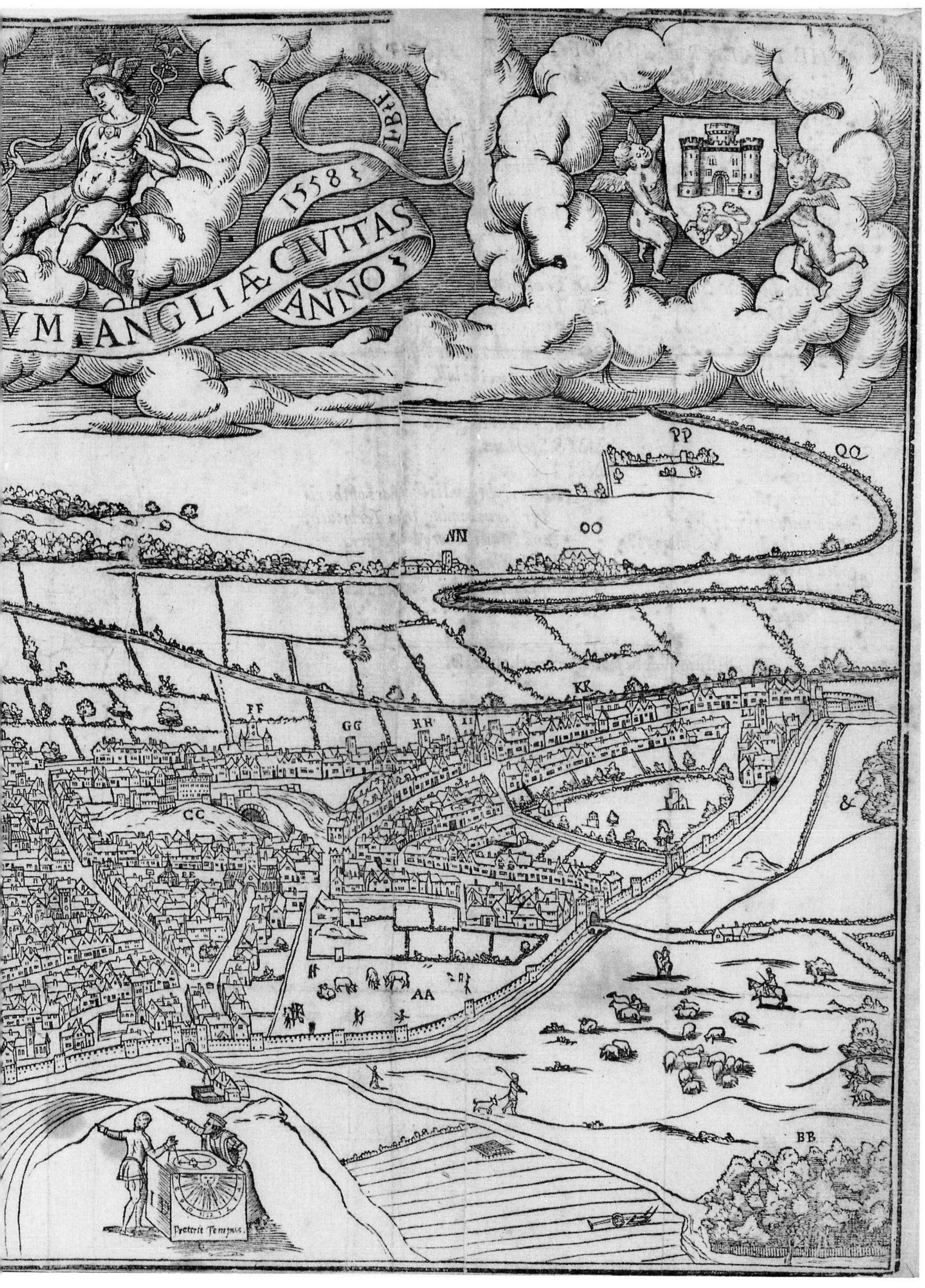
VM. ANGLIÆ CIVITAS
ANNO
1558
PP
QQ
OO
NN
KK
FF
GG
HH
II
LL
CC
EE
AA
BB
&
Preterit Tempus

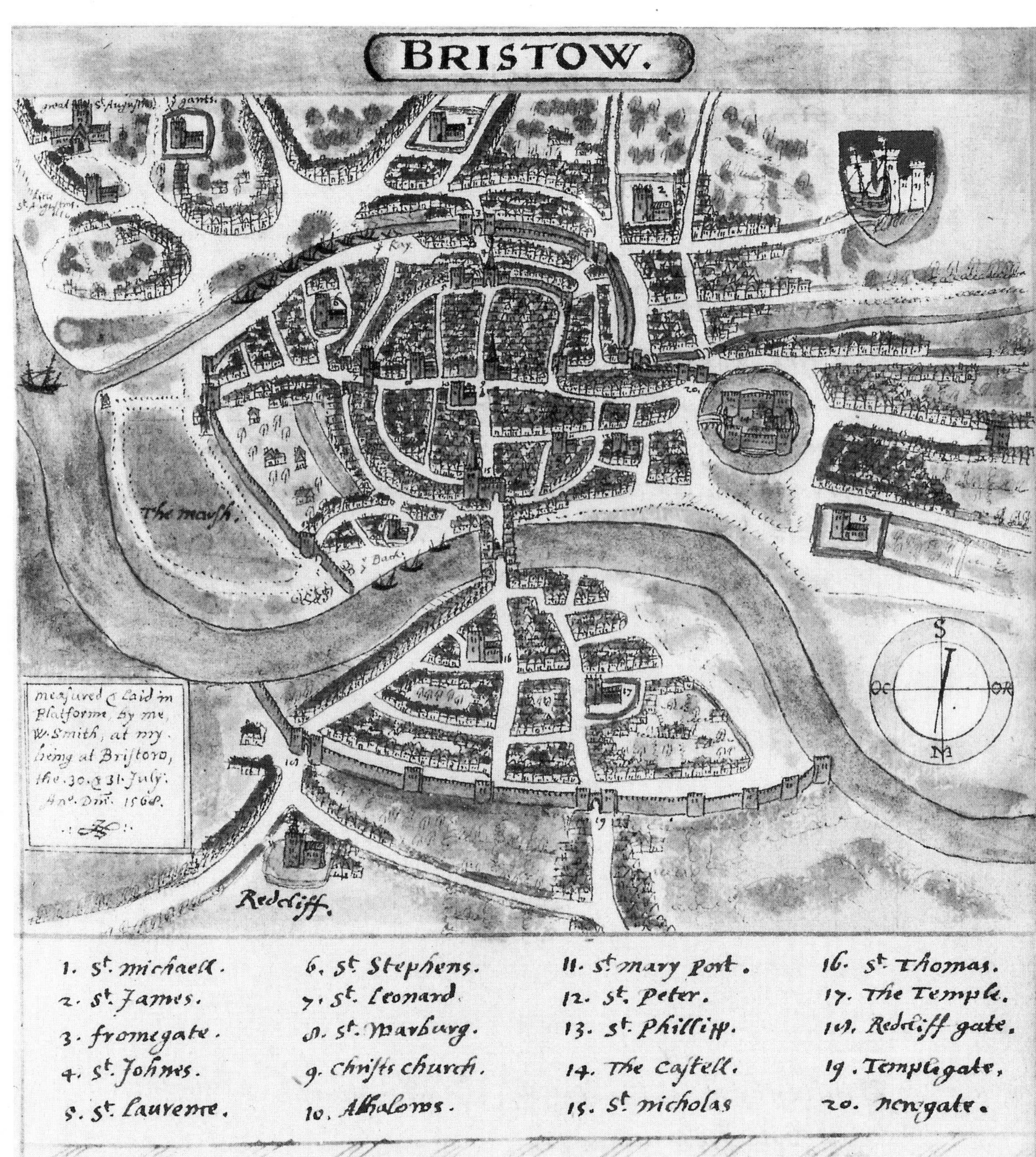

(21) Map-view of Bristol. In William Smith, 'The Particuler Description of England with the portratures of Certaine of the Cheiffest Citties and Townes 1588'.
BL Sloane MS 2596, ff.76v–77.
The map-view is oriented with North at the top, although the compass needle points to magnetic South (M = Meridies). On the next page Smith describes Bristol as 'one of the greatest and famoust [*sic*] Citties in England . . . one quarter of the Cittie standeth in Somersetshire. But the Bristollians will be a Shyre of them selves & not accompanied in any other shyre . . . There is no dunghill in all the Cittie, nor any sinck, that cometh from any howse, but all contained under the ground . . .'.

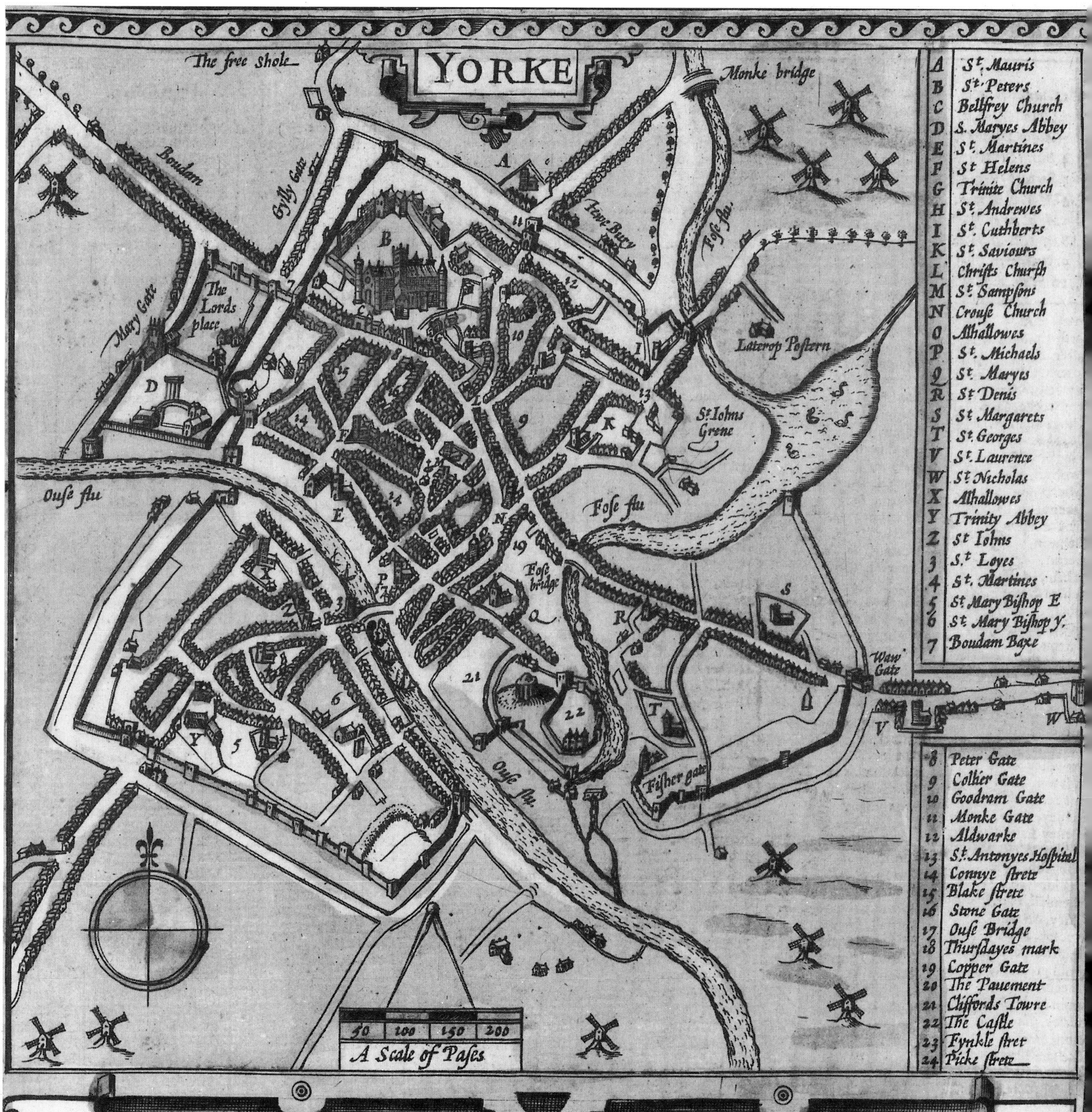

(22) 'The West Riding of Yorkeshyre with the most famous and fayre citie Yorke described'. From John Speed. *Theatre of the Empire of Great Britaine*. (London: John Sudbury & George Humble, 1611–12). BL Maps C.7.c.20. [Detail.]

by county and contained brief geographical descriptions of the major towns, forests and castles, together with general notes on history, fairs and parliamentary representation. It included eight prospects (Winchester, Colchester, Salisbury, Chester, Oxford, Stafford, Lichfield and Coventry) and seven bird's-eye views (Canterbury, Rochester, Bath, Cambridge, Bristol, Norwich and London) as well as a view of Stonehenge. While all the prospects appear to be Smith's work, only the plan of Bristol was definitely based on his own original survey.

For a printed compendium of town plans of Britain, the public had to await the appearance of *The Theatre of the Empire of Great Britaine*, 1611–12, by John Speed (**22**). The *Theatre* contained plans of over 70 of Britain's more important towns and cities as insets to the county maps. Of these, a large number are believed to have been surveyed by Speed himself. Forty-five of the plans carry the 'scale of pases' (paces) referred to in the introduction to the *Theatre* as being visited by 'mine owne travels, and into them for distinction sake, the scale of pases . . . five foote to a pase I have set'. Speed provided many British towns with their earliest known printed plans. Nine of his plans, of Cork, Dublin, Galway, Enniskillen, Lancaster, Limerick, Richmond (Yorkshire), Shrewsbury and York, subsequently appeared in book VI of the *Civitates Orbis Terrarum*, first published in 1617. Speed himself freely copied from William Smith for his plans of Bristol and Norwich, from Lyne for Cambridge, from Hooker for Exeter, from John Norden for plans of Westminster, London and Chichester, and from various other sources besides.

John Norden (*c.*1550–1625), a contemporary of William Smith, seems also to have considered the potential value of a systematic survey and mapping of the major towns of the realm. His books on English counties were intended to form part of a *Speculum Britanniae* ('Mirror of Britain'), and in his manuscript description of Northamptonshire (1591) he urged 'the most principall townes Cyties and castles within every shire should be briefly and expertly plotted in their estate and forme as at this day they are'. He failed, however, to obtain continued support for the *Speculum* and subsequently earned his living as an estate surveyor working for a number of prominent landowners, including James I and his sons Henry and Charles (later King Charles I) when they were Princes of Wales. Norden's manuscript 'Description of the Honor of Windesor', 1607 (**23**) consists of a set of plans, one of which incorporates a map-view of the town of Windsor; the detail emphasises the close links between estate survey and urban cartography.

The work of Smith and Norden also features in a cast silver medal by Nicolas Briot commemorating Charles I's return to London from his Scottish coronation in 1633. The medal bears the inscription 'As the returning sun illuminates the world, so does the King's return gladden the City', and features a view of the city broadly based on that by Smith in his 'Particuler Description' and by Norden in his printed description of Middlesex, 1593.

Manuscript plans had long been made for military purposes, and maps of strategic fortresses and ports in England survive from at least 1530. Dover had been a major concern of government since the fifteenth century, its strategic importance being matched by the destructiveness of the sea and the port's tendency to become blocked by shingle collecting at the foot of the cliffs. By 1600 the principal harbour works were concentrated to the west, near the present Hoverport. The plan of 1595 (**24**) represents one of many attempts to portray the state of the harbour and surrounding fortifications, and can be attributed to either Thomas Digges or John Hill. Digges (*c.*1543–95), a Kentish squire, was acclaimed by contemporaries as one of the leading English mathematicians of his age, and was concerned with the popularisation and practical application of the latest advances in the field. This found expression in a series of works

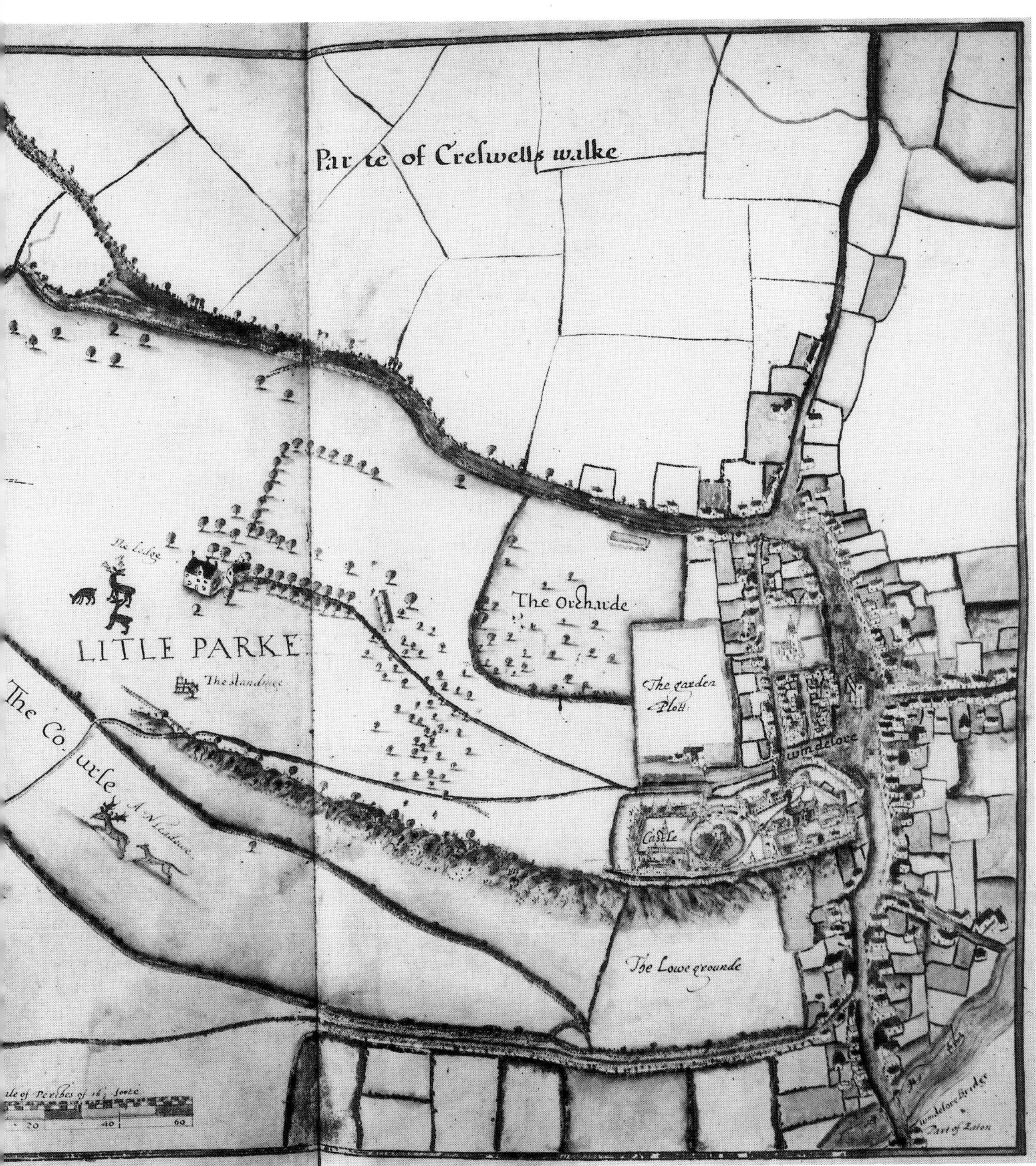

(23) Plan of the Little Park of Windsor, with Windsor Castle and the town. From John Norden. 'Description of the Honor of Windesor', 1607. BL Harley MS 3749, ff.5v–6.

on surveying as well as in maps and charts. Digges had been appointed Surveyor General of the works at Dover in 1582. By 1590 Hill was also working as surveyor there and was active as a servant of the Crown until about 1610.

Another port of strategic importance to England was Flushing (Vlissingen). Its deep-water harbour made it a haven for enemy ships, whether Spanish ships threatening England or Dutch ships mounting a commercial challenge to England overseas. At the same time it

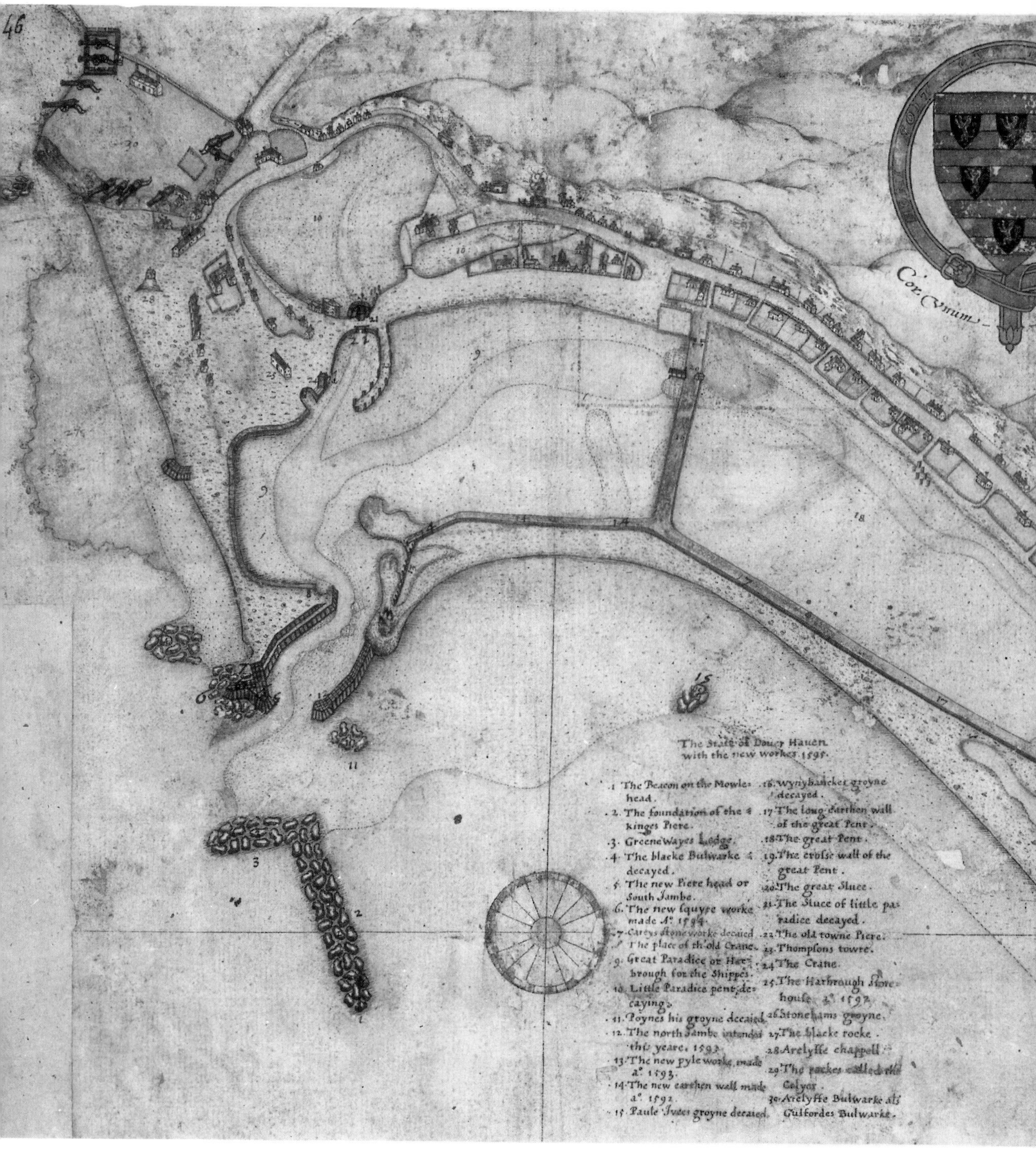

(**24**) Thomas Digges or John Hill, 'The state of Dover Haven with the new workes, 1595'.
BL Cotton MS Augustus I.i.46.
The plan shows the state of the harbour in 1595, and gives notes on recent as well as 'decaied' works in need of replacement (*eg* nos 10, 21). It was made for Elizabeth I's Lord Treasurer, Lord Burghley, who had a keen interest in maps of all kinds, and whose arms are prominently displayed.

dominated the mouth of the Scheldt and thus controlled access to Antwerp, a major Spanish-held military and commercial base. Flushing was occupied by English troops in 1585 under the terms of the Treaty of Nonsuch, as a security for the eventual repayment of English aid to the Dutch rebels against Philip II. In early 1588, however, Flushing itself was threatened by Philip's commander in the Netherlands, the Duke of Parma, and this event clearly occasioned the plan. Robert Adams, the surveyor (*d.* 1594), was a skilled architect, draughtsman, military engineer and map-maker who rose to be Surveyor of the Queen's Works. His father Clement (*d.* 1587) had been one of England's earliest native-born engravers as well as a noted scholar. The amount of decoration on the map-view, which in technique owed much to foreign example, suggests that it may have been intended for presentation to Lord Burghley or even Elizabeth I herself.

Not until the late seventeenth century, however, did printed plans become available in which accuracy as well as artistic skill was paramount. One such example was a plan-view of the city of Bristol, executed in 1673 (**25**) by James Millerd. Millerd, a mercer and Guardian of the Poor, received the gift of a silver tankard valued at £10 7*s* 6*d* from the Corporation for his plan, described as 'the largest, exactest and handsomest that ever was yet drawn of this city'. The plan was revised and reissued at least four times, in 1684, 1696, *c.*1710 and between 1728 and 1730. The rapid expansion of the city in this period acted as an incentive to Millerd to amend and reissue his map in order to maximise the recovery of his initial expenditure in a local and therefore limited market. The successive states of the plan are thus a valuable source of information to the local historian.

Millerd's map was surpassed in accuracy if not in beauty by Ogilby and Morgan's plan of London, published in 1676 as a consequence of the Great Fire ten years earlier (**26**). In order to settle boundary and property disputes and to portray the new buildings, the Lord Mayor and Court of Aldermen commissioned as 'sworn viewers', or surveyors, John Ogilby and his wife's grandson, William Morgan. In preparing the survey, Ogilby was in close consultation with Robert Hooke, Curator of Experiments of the Royal Society. Hooke suggested the scale (100 feet to the inch) and advised on surveying and cartographic techniques. Ogilby was a former dancing teacher, Master of the Kings Revels at the Restoration of Charles II, and the King's 'Cosmographer Royal and Geographik Printer', who had lost his house, shop and entire stock in the fire. The result of his efforts was London's first true 'plan' – that is, a map on which all the detail, including streets and houses, is drawn to scale. Published in 20 sheets, it covered the area from Moorfields to the Thames and from Holborn to Whitechapel, and was accompanied by a booklet entitled 'London survey'd: or, an explanation of the large map of London', which contained an outline of the city's history and institutions, a table of streets and a description of the conventional signs used on the map. The 'explanation' also carried an advertisement, for both the map of London and Ogilby's *English Atlas*, which cautions the reader that 'because several counterfeit Books and Maps, notoriously false, especially of London, have been and others are Preparing to be Publish'd, you are Desir'd to Receive no Book or Map for Part of the Atlas or Survey, that hath not the Names of John Ogilby or William Morgan, or both'.

(**25**) James Millerd. 'The famous city of Bristoll' (Bristol: James Millerd, *c*.1710.)
BL Maps K.Top XXXVII.32.

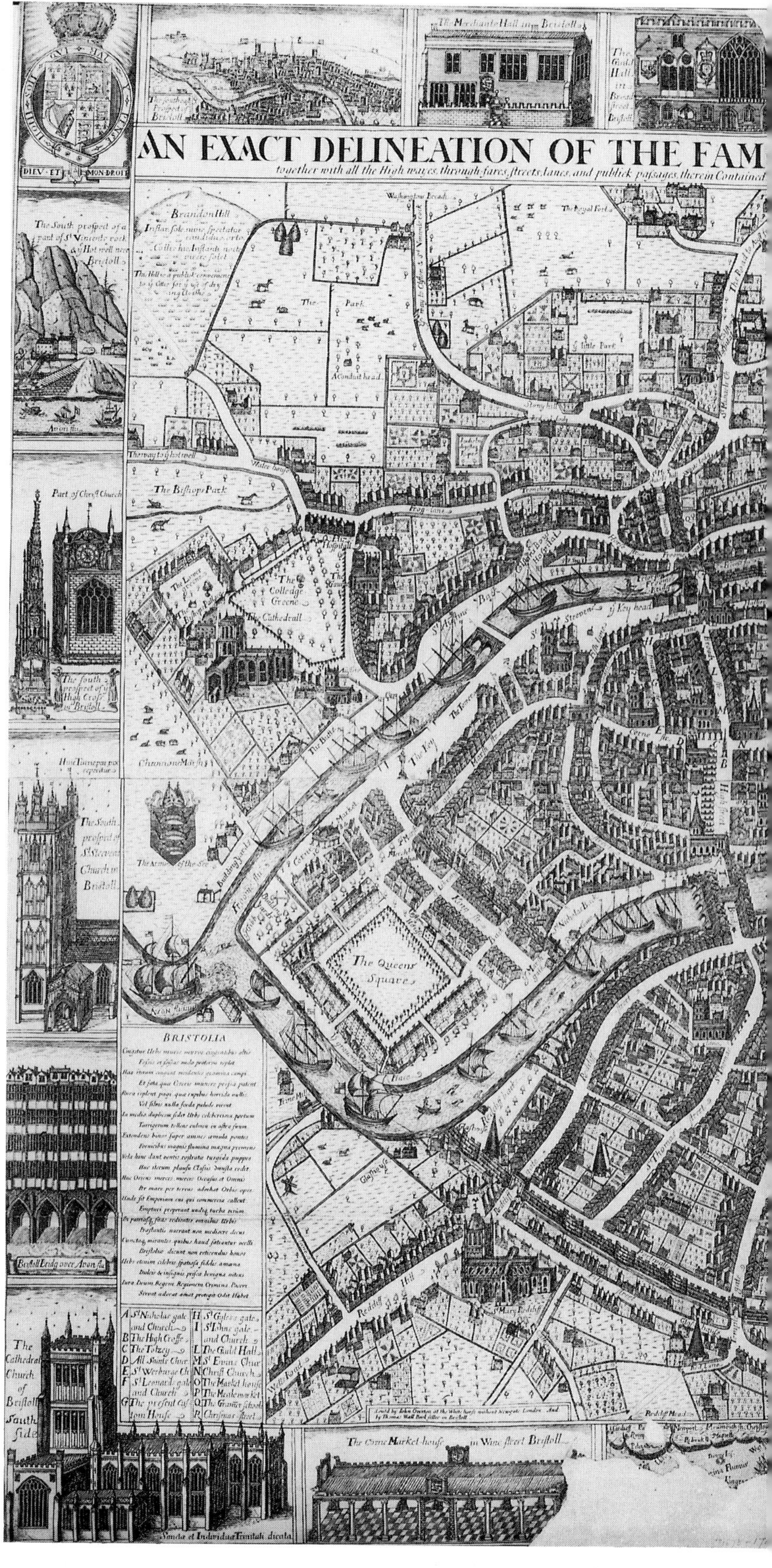

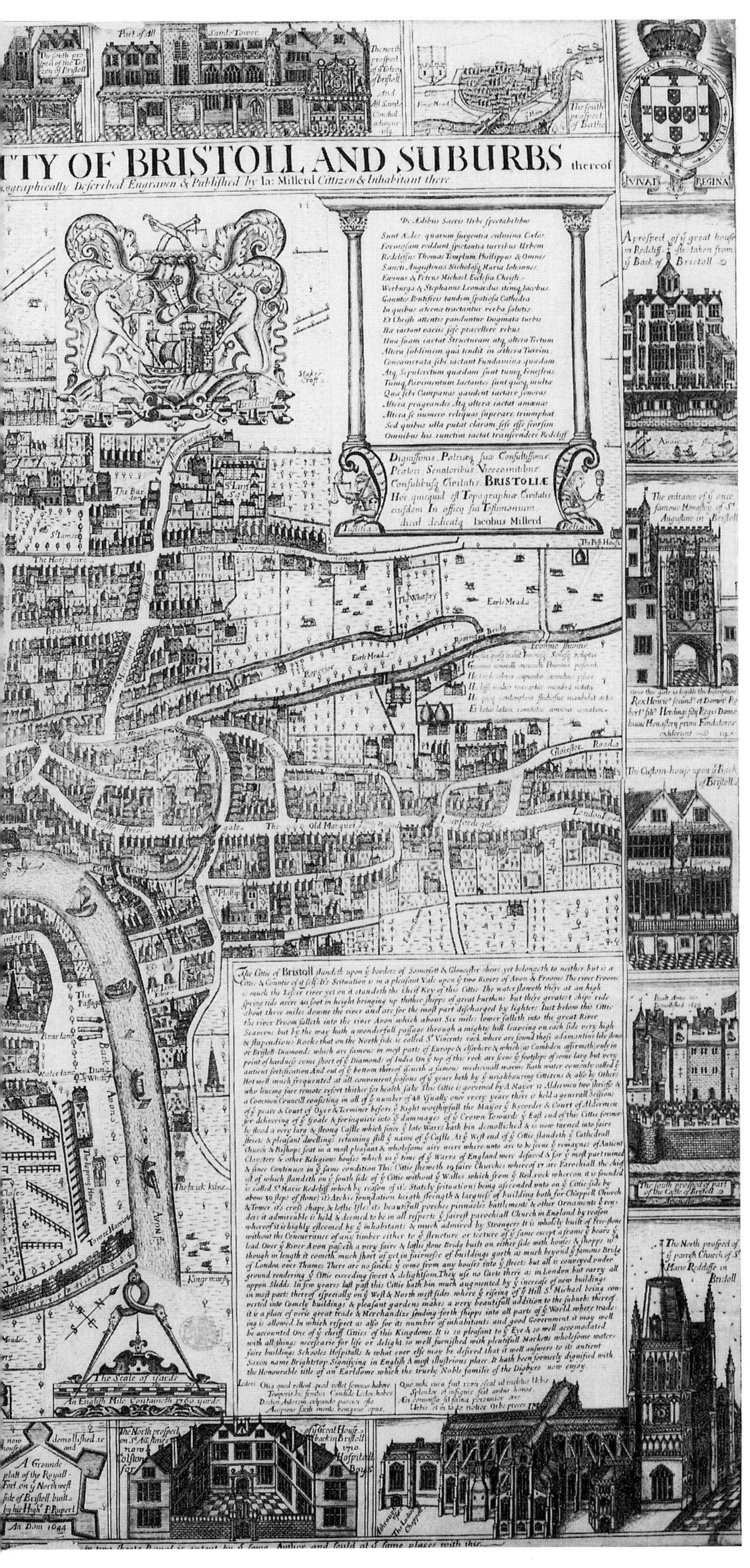
TY OF BRISTOLL AND SUBURBS thereof
ographically Described Engraven & Published by Ia: Millerd Cittizen & Inhabitant there
De Ædibus Sacris Urbe spectabilibus
VIVAT REGINA
Dignissimis Patriæq suæ Consultissimis Prætori Senatoribus Vicecomitibus Consulibusq Civitatis BRISTOLIÆ Hoc quicquid est Topographiæ Civitatis eiusdem In officij sui Testimonium dicat dedicatq Iacobus Millerd
Earls Mead
The Horse faire
Broad Mead
Old Marquet
Glocester Road
Avon
The brick kilns
Kings marsh
The Scale of yards
An English Mile Containeth 1760 yards
A prospect of ye great house in Redcliff-strt taken from ye Back of Bristoll
The entrance of ye once famous Monastery of St. Augustine in Bristoll
The Custom-house upon ye Back of Bristoll
The south prospect of part of the Castle of Bristoll
The North prospect of ye parish Church of St. Marie Redcliffe in Bristoll
A Grounde platt of the Royall Fort on ye Northwest side of Bristoll built by his Highs P. Rupert
An Dom 1644

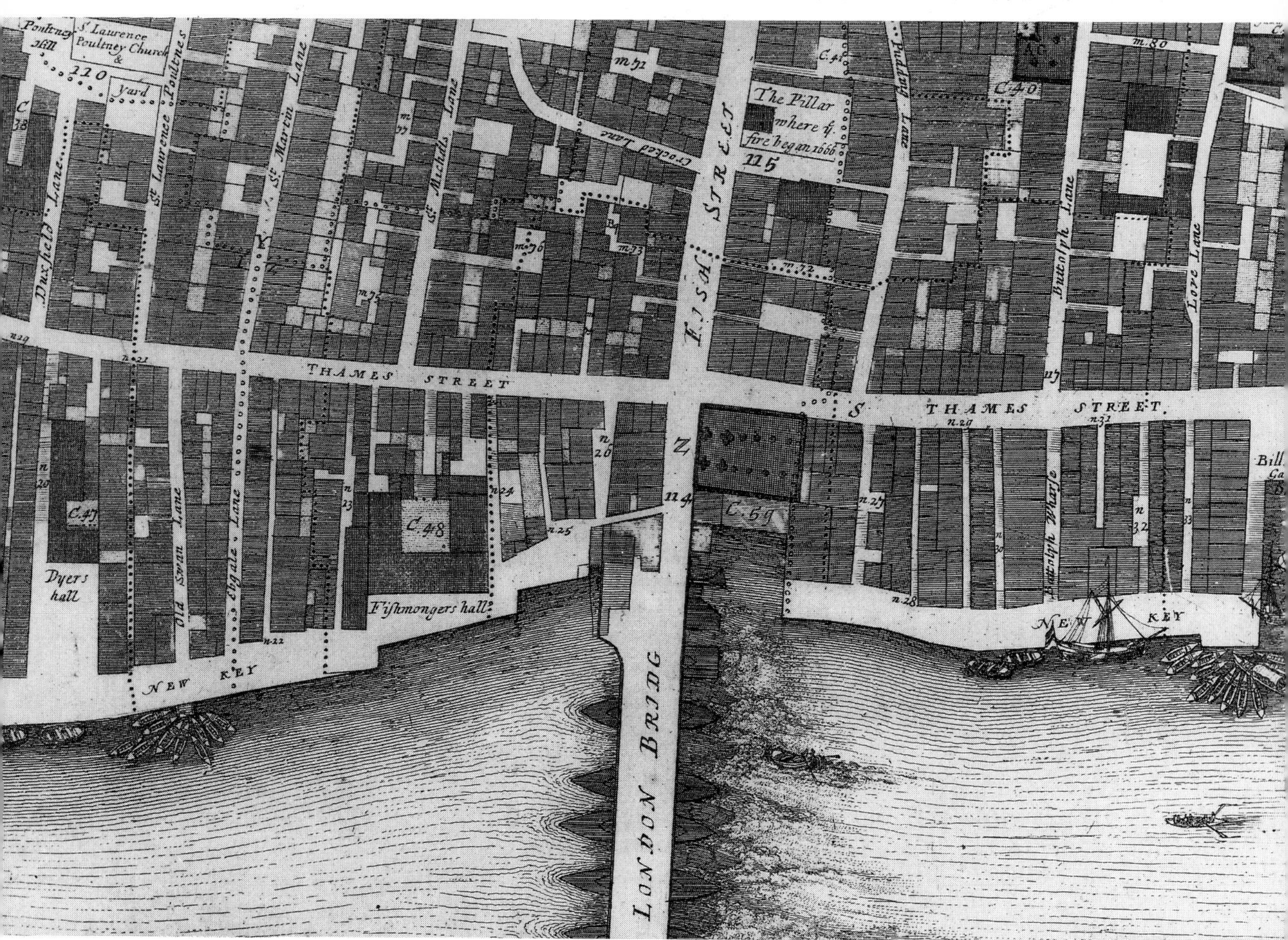

(**26**) 'A large and accurate map of London . . . Made by John Ogilby and William Morgan'. (London, 1676.) BL Maps Crace II.61. [Detail.] The plan bears witness to the speed with which London had recovered from the fire, for by 1676 the city had largely been rebuilt. The Monument, marked as 'The pillar where ye fire began 1666', then in its final stages of completion, can be seen on the east side of Fish Street as it runs north from London Bridge. The interior ground plans of the rebuilt churches are also shown, as are the vacant sites of other churches destroyed by the fire.

4

The baroque city: the town plans of the eighteenth century

The town plans of the eighteenth century reveal further developments in urban cartography and they also reflect the new urban forms that were by then emerging. In Europe, increasingly powerful and centralised governments had the resources necessary to impress themselves physically on the design and structure of the city. Berlin, for example, had been a major European capital since 1709, when the new suburbs of Dorotheenstadt and Friedrichstadt were joined with the historic core of Berlin-Kölln. King Frederick William I of Prussia (1713–40) enlarged the army and made the city a major garrison, its great new squares and avenues serving as urban parade-grounds for a military establishment which by 1740 comprised almost one quarter of the 90,000 inhabitants. One map of the city was delineated by Matthäus Seutter in about 1720, a date suggested by the presence on the map of the newly-built Quarre Market and Rondel Market. Seutter, pupil of Homann and geographer to the Holy Roman Emperor, produced numerous atlases, maps and globes as well as plans of over 90 towns and cities in northern and central Europe. All of them incorporated a prospect and an elaborate cartouche, which in the case of Berlin carried a portrait of Frederick William.

Smaller in scale, but even more dramatic in concept, was the city of Karlsruhe, founded in 1715 by the Margrave Karl Wilhelm of Baden Durlach (**27**). Although Karlsruhe was originally intended only as a palace in the Harzwald, the Margrave's ideal for a new Residenzstadt (capital) resulted in a perfect specimen of the formal and geometric town planning of the age. Blending contemporary principles of landscape gardening with the ideas of the Italian military engineers demonstrated by Braun & Hogenberg's map-view of Palmanova, the palace acted as a focus for 32 radiating streets within a concentric boulevard. Nine of the streets, to the south of the palace, made up the area of the new town, with the major commercial artery, the Langestrasse, formed by another street tangential to the circle. The plan of 1739 was drawn by a military surveyor, Johann Jacob Baumeister, and published in Nuremberg by Homann's Heirs, the printing firm that had after 1730 taken over the map and atlas business founded by Johann Baptist Homann in 1702.

St Petersburg (**28**) was another new city of the Baroque era, owing its creation to the grandiose vision and relentless energy of Czar Peter the Great. In establishing a major fortress and naval base on the Baltic as a defence against Sweden, Peter's ambitions were partly military. However, the formidable building project which began in 1703 on a group of marshy islands in the Neva River had a much wider political objective. This was no less than the creation of a new capital city which would act

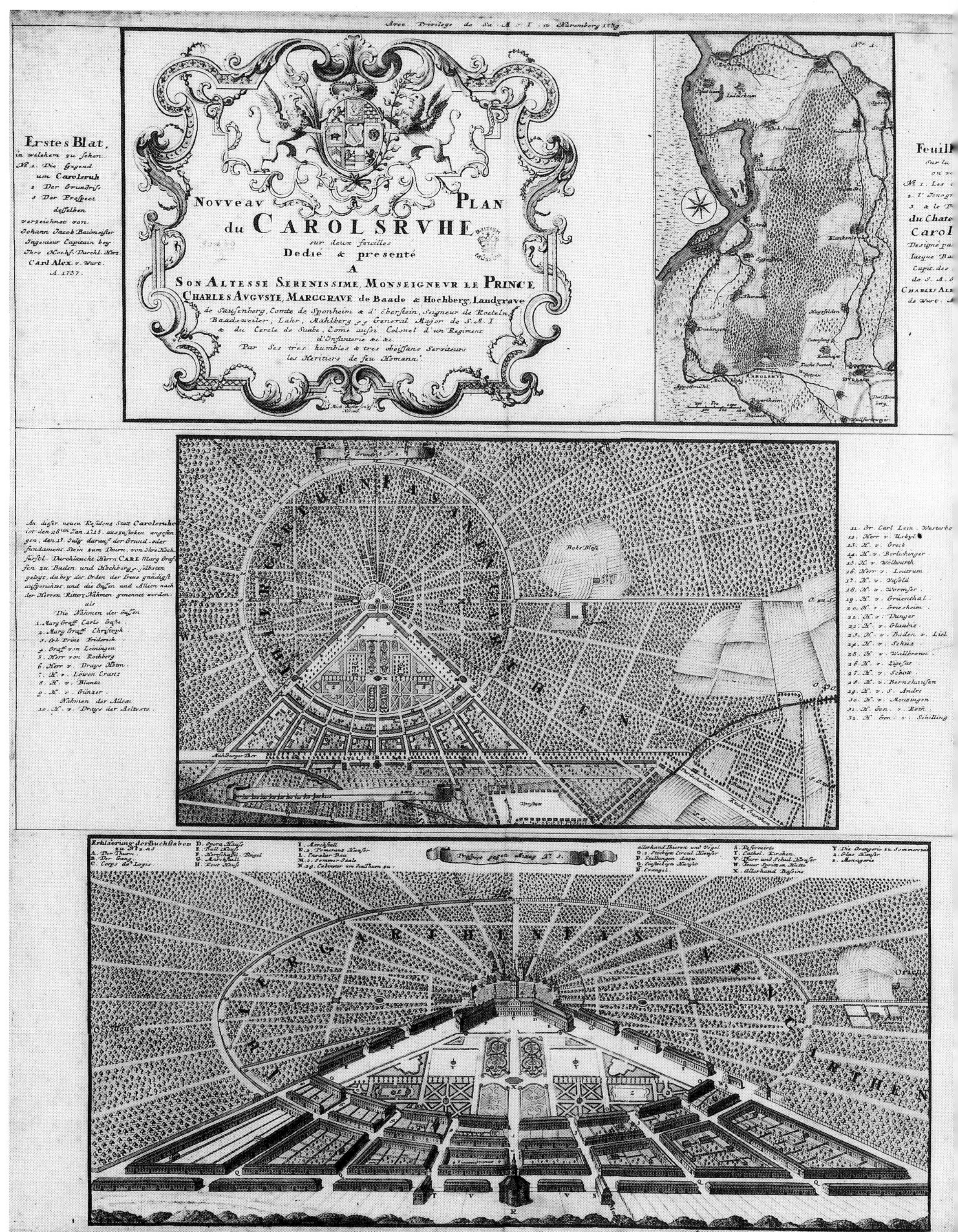

(27) Johann Jacob Baumeister, 'Nouveau Plan du Carolsruhe.' (Nuremberg: Homann's Heirs, 1739.) BL Maps 30430.(2).

(PLATE 7) Thomas Yeakell and William Gardner, 'Brighthelmston' (Brighton). (Brighton, Richard Thomas, 1779.) BL Maps K.Top. XLII.15.

Yeakell and Gardner's plan depicted a small fishing village that was already becoming popular with those who had read *Dr. Russels Dissertation concerning the Use of Sea Water in Diseases of the Glands*, 1753. This book recommended the qualities of the sea and air at Brighton. The village expanded rapidly into a fashionable resort after the first visit of the Prince of Wales in 1783. 'Till lately it had the name of Brighthelmstone; but, like low persons rising to eminence, who are often ashamed of their origins it has now assumed the title of Brighton, which certainly has a more genteel sound'. (Anon, *A Guide to all the Watering and Sea-bathing*, 1810).

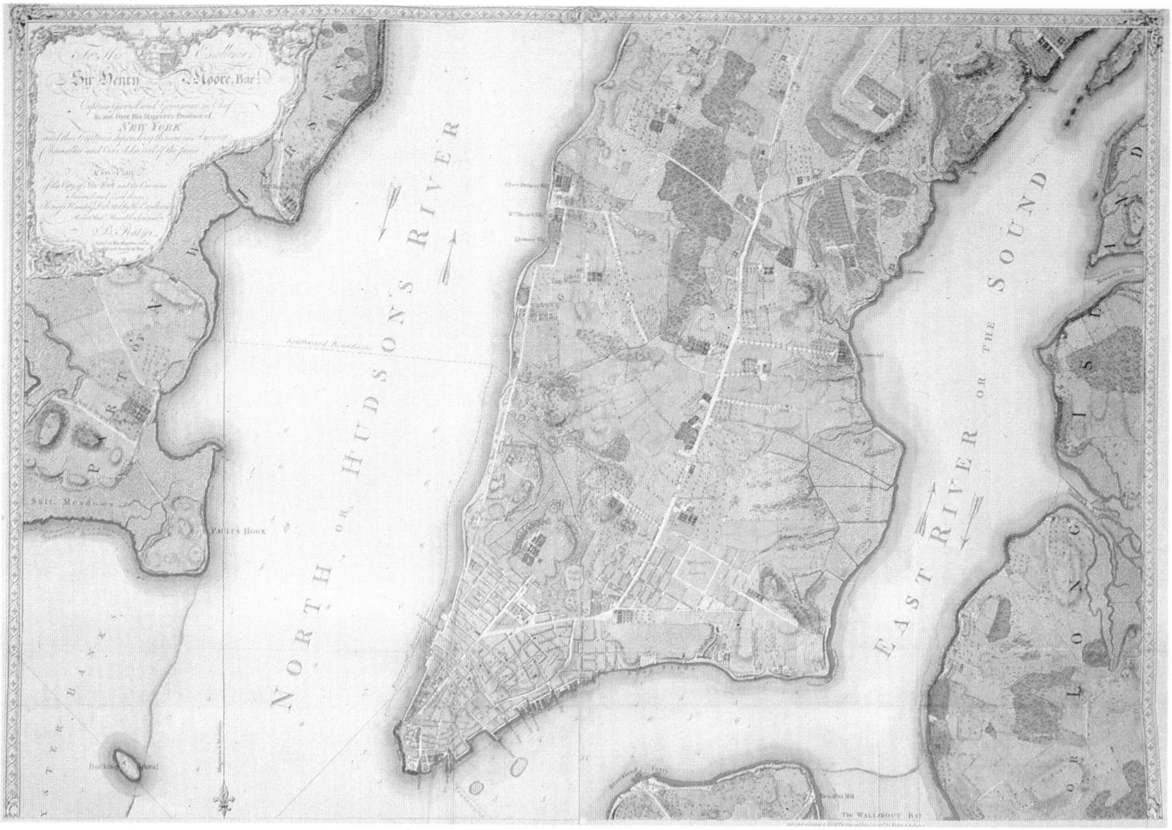
Sir Henry Moore, Bar!
NEW YORK
NORTH OR HUDSON'S RIVER
EAST RIVER OR THE SOUND
Salt Meadow
PAULUS HOOK
The WALLABOUT BAY

THE MARKET PLACE TAKEN FROM NEAR THE POST OFFICE

(PLATE 8) Bernard Ratzer, 'Plan of the city of New York', north sheet. (London: Jeffreys and Faden, 1776.)
BL Maps K.1.Tab.44.
New York is shown to extend as far north as the site of Times Square. Beyond are large tracts of open country occupied by estates, including those of the Stuyvesant family, which later gave their names to places in the modern city.

(PLATE 9) Thomas Hornor, *Plan of the town and parish of Kingston upon Thames.* (London: Thomas Hornor, 1813.)
BL Maps K.Top. XL.15–3.
Hornor's plan of Kingston upon Thames, in his own words, 'comprehends the town on a large scale, shewing in a distinct manner the various particulars of the streets . . . with a reduced plan of the parish . . . and a view of the market-place; the whole, blended into one design by a picturesque fore-ground, forms a faithful view of the parish'.

as a 'window onto Europe'. From here, Peter hoped to turn the Russian empire away from its roots in the east towards the culturally and economically more sophisticated climate of western Europe to which the Baltic gave access.

The 'Plan de la ville de St Petersbourg' was commissioned in 1753 by the Geography Department of the Imperial Academy of Sciences to celebrate the city's 50th anniversary. The Academy was also one of Peter's creations, and his policy was reflected in the selection of the French geographer and astronomer Joseph Nicolas Delisle to fill the first chair of the Department of Geography. He held this post until 1747, and the continuing French influence is apparent in the title of the plan. The cartographer, Ivan Fomich Truskott (John, son of Thomas Truscott), was evidently of British descent. Copies of the plan were distributed to the major European capitals to inform a wider audience of the new city's splendours.

In Asia and America, new urban forms, transplanted from Europe, determined the shape of the cities of the colonial empires. In the 'View of the town of Savannah as it stood the 29th of March 1734' (**29**), the gridiron, or rectilinear street plan, can be clearly seen. This egalitarian and cost-effective style of urban planning, originating in the new squares of London and Berlin, was to become characteristic of the cities of North America. The view was executed by Peter Gordon, the first bailiff of the city, Conservator of the Peace, and one of the 120 colonists who had set sail from Deptford in October 1732. The majority of the colonists were convicts, personally selected and supervised by James Edward Oglethorpe (1696–1785), penal reformer and philanthropist and a leading figure of the Trustees for Establishing the Coloney of Georgia in America. The view, designed to show the progress made in a remarkable experiment in social engineering, was, appropriately, dedicated to the trustees.

The older colonial foundations were by this time emerging as considerable cities in their own right, as demonstrated by the map-view of Manila, prepared for Philip V of Spain by the governor of the Philippines, Don Fernando Valdez Tamon. In 1739 he was appointed Mariscal de Campo (Field Marshal) and recalled to Spain, the map forming part of his final report. He intended to present a copy to his king in person, but died *en route* to his native country. The map revealed a city typical of the Spanish, Portuguese, Dutch, French and English overseas empires at any time from the seventeenth to the twentieth centuries. The Spanish citadel, established in 1571, formed the administrative, commercial and military core, around which were clustered the native Filipino settlements, each with their own churches and market-places.

Contemporary with the view of Savannah and the map-view of Manila is a bird's-eye view of a much older city, also prepared as a result of an official survey. The plan of Paris, drawn and surveyed in 1739 by Louis Bretez (**30**), demonstrated the importance to a surveyor of a wealthy and influential patron. His work was commissioned by Michel Etienne Turgot, Prévôt de Marchands of Paris and effectively the city's chief administrator, who hired Bretez in preference to L'Abbé de la Grive, the official city surveyor. Ordered to complete the task in two years, Bretez was paid the relatively large sum of 10,000 livres, receiving 200 livres a month and the balance when the project was finished. A permit

(**28**) 'Plan de la ville de St Petersbourg'. (St Petersburg: Imperial Academy of Sciences, 1753.)
BL Maps K.4.Tab.44.
The island fortress of Petrapavlovsk appears centre left, whilst on the south bank of the Neva River (to the top of the map) can be seen the naval shipyards and the Admiralty building, its gilded spire the focus of the wide boulevard of Nevsky Prospect. Just left of the Admiralty can be seen the Winter Palace.

issued by Turgot entitled Bretez to enter every building in the city for the purpose of producing the numerous individual sketches from which the plan was composed.

When published in 1739, the 'Plan de Turgot', as it came to be known, aroused little comment in the Parisian press, despite the copies that were sent free to the leading figures of the court and to every French ambassador. It is possible that Bretez's finished work, a bird's-eye view on 20 sheets, no longer matched the standards of accuracy by then expected of urban surveyors, who by the 1730s were commonly depicting towns by means of the less decorative but more precise plan.

Propaganda of a more overtly political nature was also present in a bronze medal made in Rome in 1721 by Otto Hamerani entitled 'Appeal against the House of Hanover'. Depictions of towns – usually derived from printed views – on coins and medals were common between 1650

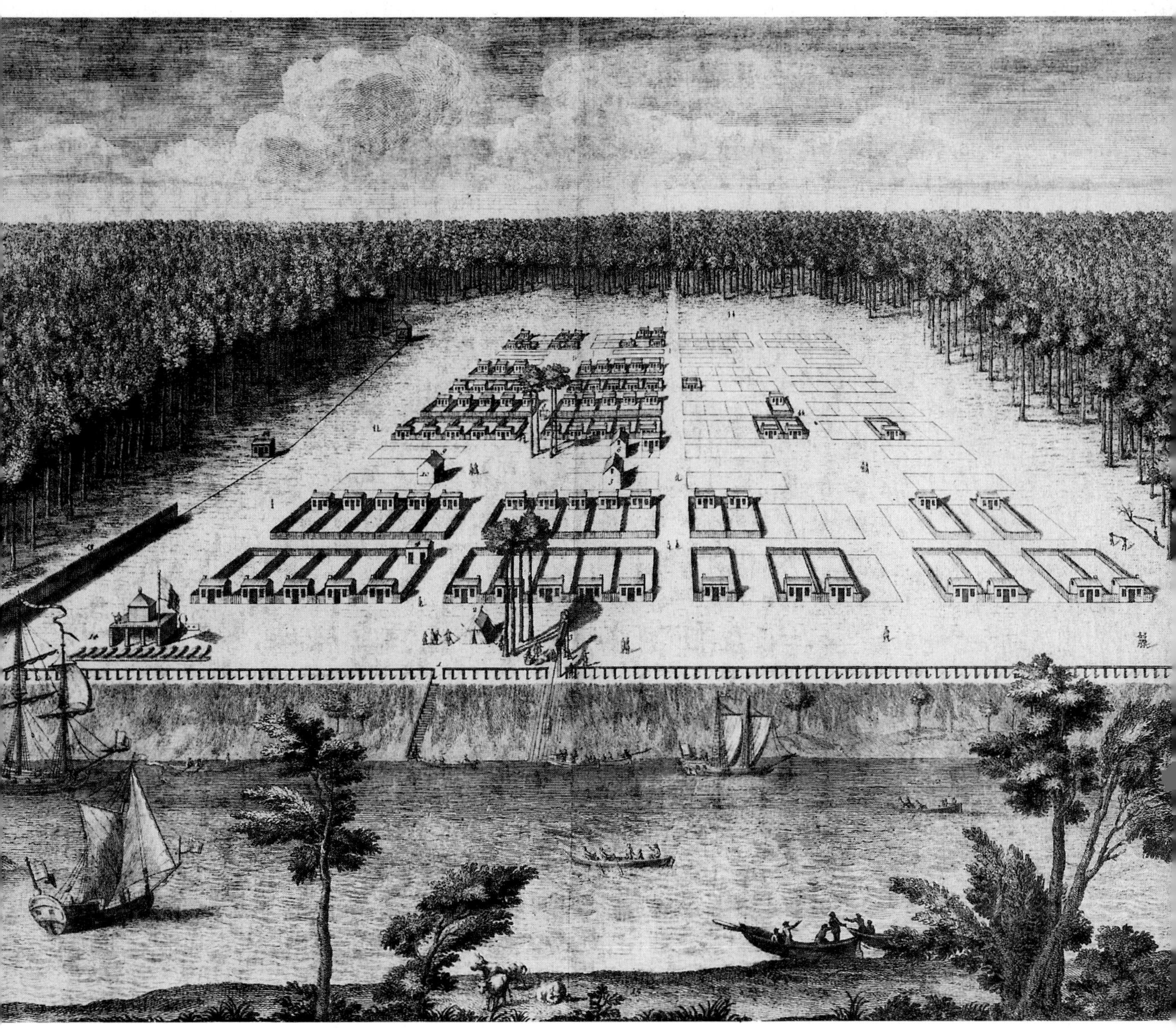

(**29**) Peter Gordon, 'A view of the town of Savannah as it stood the 29th March 1734'. (London? 1734.) BL Maps K.1.Tab.44,f.43.

and 1750, particularly in Germany and Italy. The medal formed part of an attempt to link George I and his dynasty with the political and financial chaos which followed the collapse of the English South Sea Company in the autumn of 1720.

In Britain, as in Europe generally, the eighteenth century witnessed a great improvement in surveying methods, and town plans were correspondingly expected to be produced to a higher standard of accuracy. As the historian William Maitland complained in the preface to his *Survey of London* of 1739:

> I once intended to annex a modern Plan of London to this Work; but as the Plans, from which I must have taken the same, are all very defective, and perhaps the worst of the Kind extant, the best of them would rather be a Disgrace than an Embellishment of the Work: I therefore thought fit to omit it, as it could be but of little Use to the Reader.

Cloître N.D.
Parvis N.D.
HOTEL DIEU
Petit Chatelet
Petit Pont
S.T SEVERIN
Col. de M.e Gervais
MARCHÉ NEUF
PLACE
Rue de la Juiverie
R. de la Barillerie
Cour du Palais
PALAIS
S.t Michel
Cour du Palais
Horloge du Palais
Conciergerie
Cour neuve du Palais
Quai de l'Horloge
Quai des Morfondus
Quai des Orfevres
PLACE D'AUPHINE
Quai des Augustins
LA VALLÉE
LES GRANDS AUGUSTINS
Rue Dauphine
R. de Nevers
Pont
neuf
Statue d'Henri IV
Samaritaine
de la Ferraille ou de la Megisserie
Fort l'Evêque
Carrefour des Trois Maries
Rue Thibautaudé
de la Monnoie
Quai de l'Ecole
Quai Bourbon
Cloître S.t Germain
S. GERMAIN L'AUXEROIS
Quai de Conti
H. de Conti
Col. des Quatre Nations

(30) 'Plan de Paris . . . dessiné et gravé sous les ordres de Michel Etienne Turgot'. (Paris, 1739.)
BL Maps C.11.d.10.

Stimulated by the public demand, urban cartographers produced works which became progressively more accurate and more functional, and by mid-century the decorative bird's-eye view and map-view forms had been largely abandoned except for the older and more picturesque towns. Typical of the high standards of the large-scale plans of its day was Thomas Hanson's Birmingham, 1778 (31), the third complete survey to be carried out in the city in 50 years. When compared with the plans by Westrey (1731) and Bradford (1756), Hanson's survey shows the dramatic growth of the city in this period, from a market town of 15,000 inhabitants in 1731 to a major industrial centre of 42,000 by 1778.

In London, general topographical mapping was supplemented by maps of wards (32), the administrative units into which the City had been divided since Norman times. There were eventually 26 wards, each electing an alderman (normally for life) and common councilmen, who were responsible for street cleaning, policing, lighting and other local government duties. The ward also served as a convenient unit of survey. John Strype's *Survey of the Cities of London and Westminster*, 1720, comprising a revised edition of Stow's survey of 1598, was the first to contain a series of such maps and they continued to be published, either individually or, more commonly, in series, until the 1840s.

While the plan depicted the horizontal dimensions, the prospect, or view, provided a vertical complement. The most successful and prolific publishers of prospects in the first half of the eighteenth century were Samuel and Nathaniel Buck (33), who depicted 76 different towns and dockyards in the years 1721 to 1753. They also published a county-by-county survey of the nation's antiquities.

The most famous urban cartographer of the century was John Rocque, a Frenchman who settled in England and who published over 100 maps, plans and road books between 1734 and his death in 1762. His town plans were surveyed by trigonometrical observations from towers and other tall buildings and by checking the results with instrumental measurements of angles and distances taken on the ground. In the plan of Exeter of 1744 (34) his portrayal of the surrounding market gardens, orchards and pasture betrays his earlier career as an estate surveyor and 'dessinateur de jardins'. His prolific output provided the basis for a number of works published after his death. His brother-in-law, Andrew Dury, had already published maps in co-operation with Rocque's widow, Mary Anne, by the time he issued *A Collection of Plans of the principal Cities of Great Britain and Ireland* in 1766 (35). This pocket atlas included maps of the coast of Great Britain and Ireland, as well as miniature plans of Exeter, Bristol, Shrewsbury, Bath, York and Aberdeen, among others. Most of these were copied from Rocque either as reductions of his large-scale plans or from insets on his county maps. The publisher, land surveyor and geographer John Andrews also derived a number of his maps from Rocque's work when he compiled *A Collection of Plans of the Capital Cities of Europe* (1771). This contained 42 plans of capital cities from all over the world, all copied from a variety of British and French sources.

The establishment by the Board of Ordnance of the trigonometrical survey was to have profound significance for the urban mapping of the

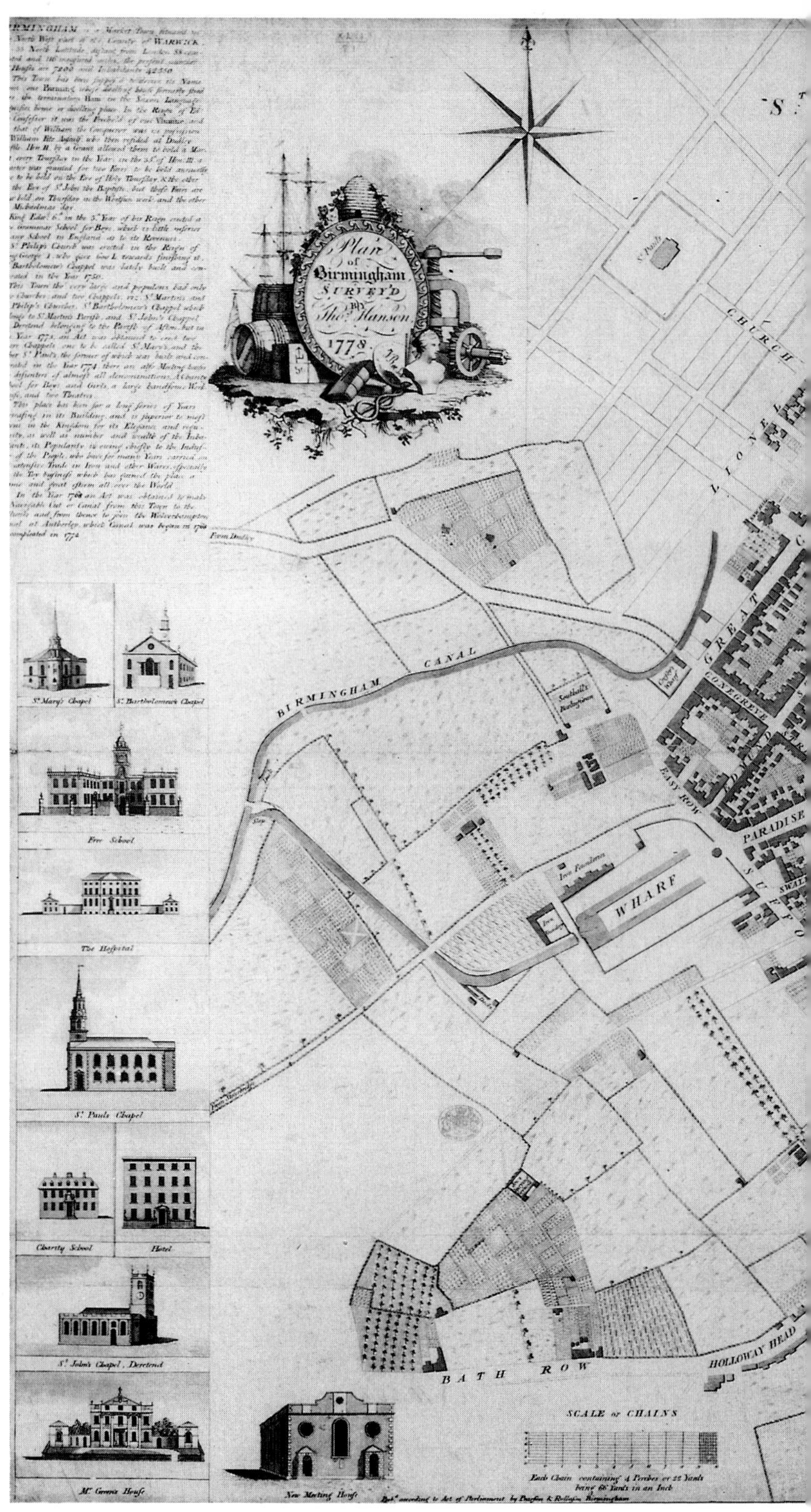

(31) 'A plan of Birmingham survey'd by Thomas Hanson'. (Birmingham: Pearson & Rollason, 1778.) BL Maps K.Top XLII.81.

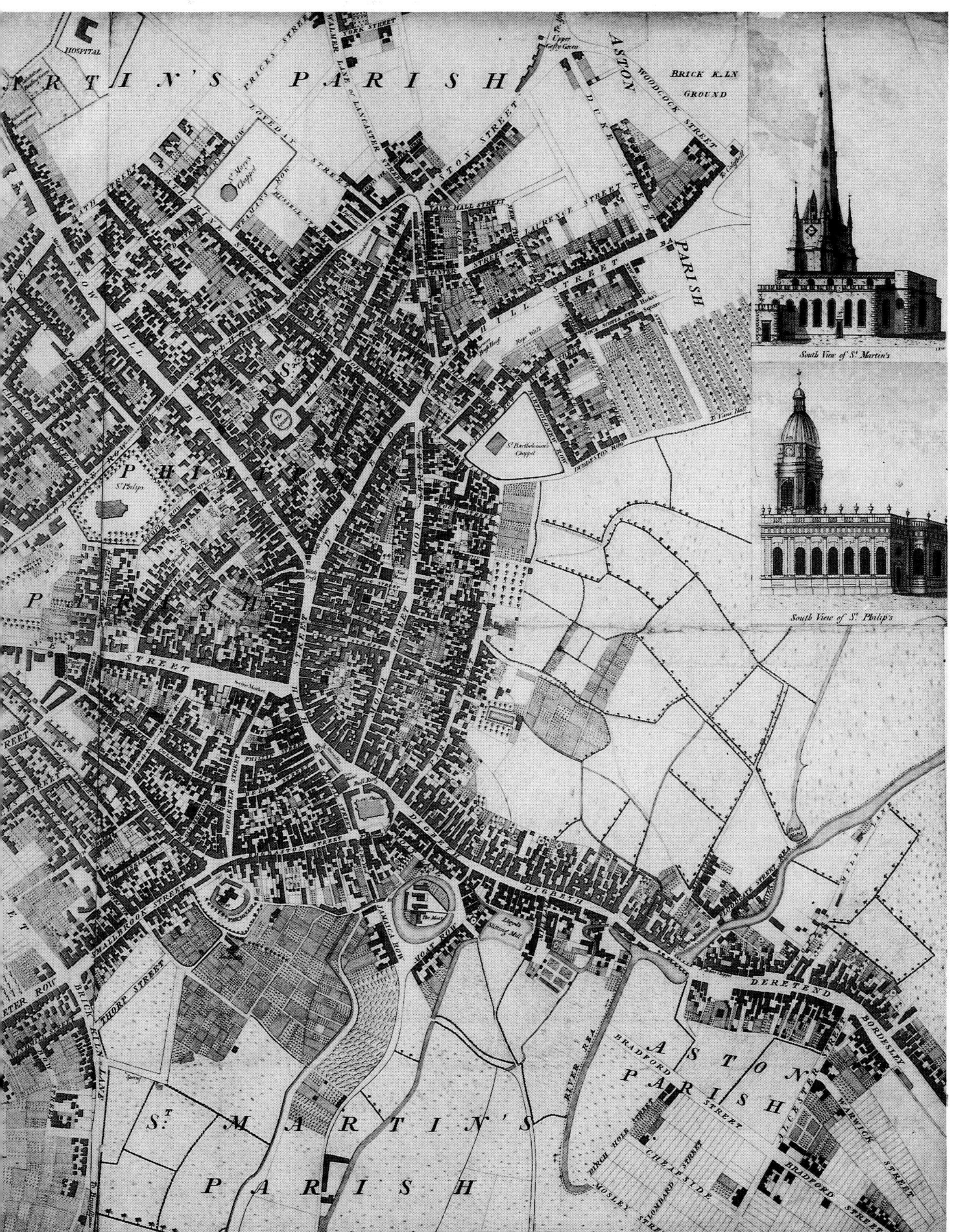

South View of St. Martin's
South View of St. Philip's

British Isles in the next century. Two early recruits were Thomas Yeakell and William Gardner, both of whom worked as salaried surveyors for Charles Lennox, third Duke of Richmond, who was expanding and developing his large estates around Goodwood, Sussex. The maps produced as a result of their activities included estate plans, town plans and the 2 inch to 1 mile 'Great Survey' of Sussex. When their employer was appointed Master-General of the Board of Ordnance in 1782, he secured the appointment of Yeakell and Gardner to undertake the land survey that he was establishing on a national basis. Their plan of Brighthelmstone (Brighton) of 1779 (PLATE 7) depicts a small fishing village then on the point of becoming a fashionable seaside resort. It can be compared with Benjamin Donne's 'A new and correct plan of the City of Bath' of 1810 (**36**), by then England's premier 'spa' and already renowned for its Georgian townscape formed by the extensions of Gay Street, The Circus, Royal Crescent and Lansdown Crescent.

The new responsibility of the Board of Ordnance underlined the close links between cartography and warfare. The years leading up to the American War of Independence witnessed much activity by the surveyors of the British Army. One of these, Lieutenant Bernard Ratzer of the Royal American Regiment of Foot, surveyed the city of New York and its environs in 1766 and 1767, issuing a plan and view of the city three years later. In 1776 the 'Ratzer' plan (PLATE 8) was reissued in response to the growing interest in the political events then taking place in the American colonies.

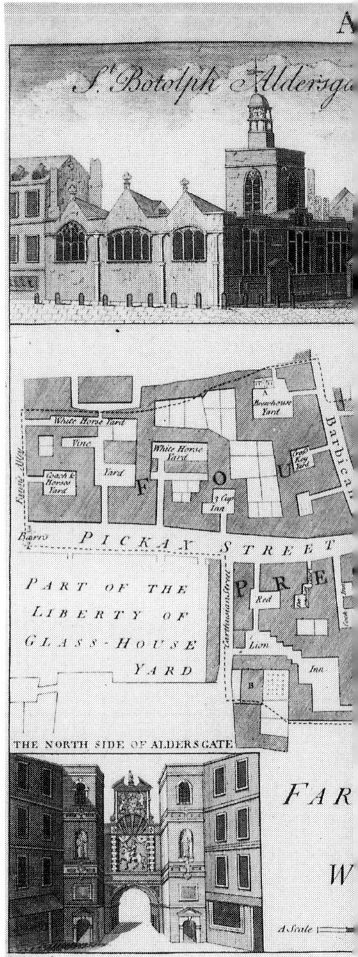

(**32**) Richard Seale, 'A plan of the ward of Aldersgate'. London: Jacob Ilive, 1739.)
BL Maps Crace viii.(2).
The plan of Aldersgate displays two features characteristic of London ward maps: the dedication to the Alderman, and the inset views of prominent buildings.

(**33**) 'The south west prospect of the University, and City of Oxford'. (From Sir Hans Sloane's collection of 24 views from Buck's *Antiquities*.) (London: Samuel and Nathaniel Buck, 1731.)
BL Maps C.10.a.21.
The view of Oxford, published in 1731, shows 'a noble flourishing city, so posses'd of all that can contribute to make the residence of the scholars easy and comfortable that no spot of ground in England goes beyond it . . . the most noble figure of any city of its bigness in Europe' (Daniel Defoe, *A tour thro' the whole islands of Great Britain*, 1724–26).

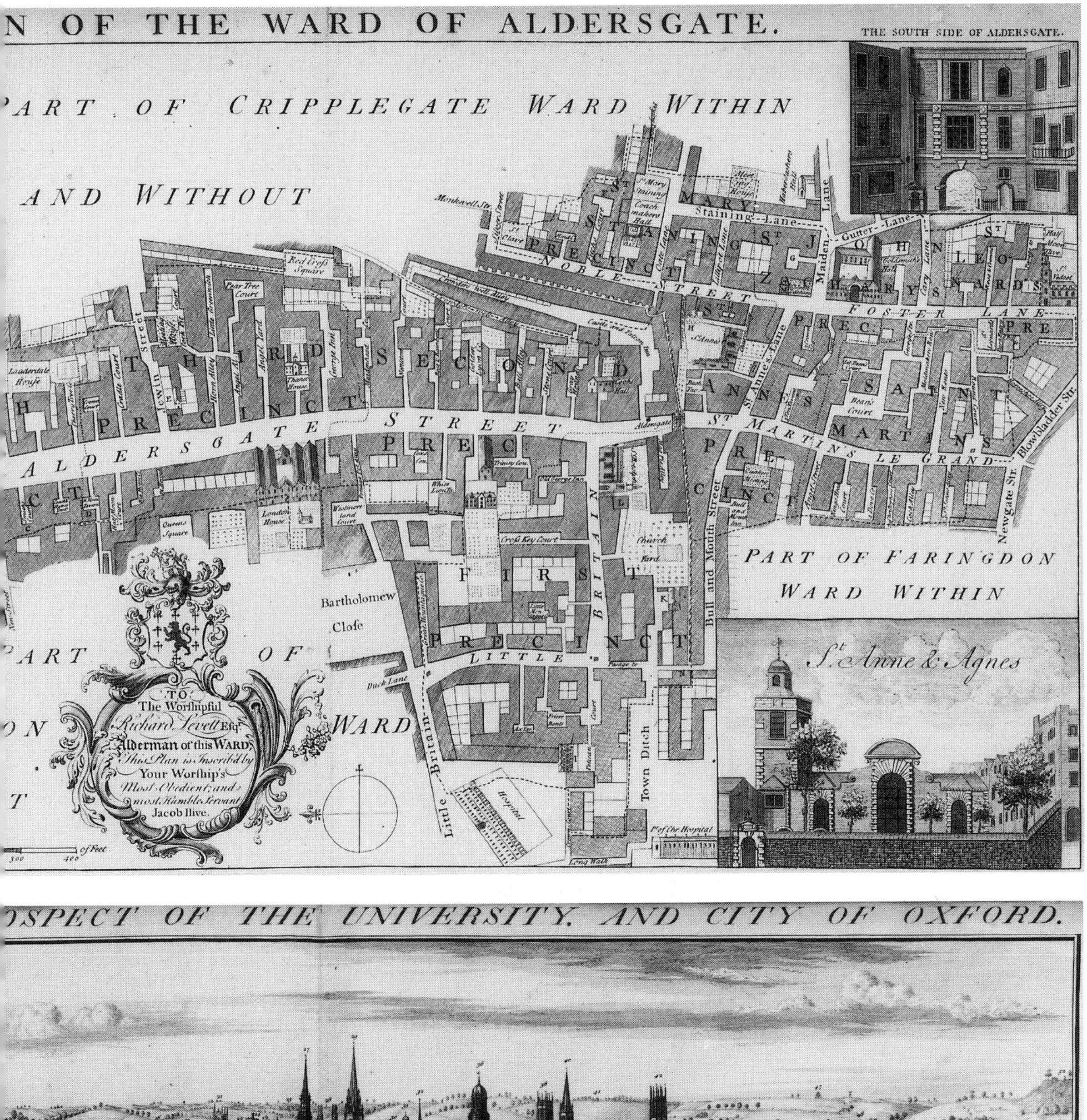

N OF THE WARD OF ALDERSGATE.

THE SOUTH SIDE OF ALDERSGATE.

PART OF CRIPPLEGATE WARD WITHIN AND WITHOUT

PART OF FARINGDON WARD WITHIN

PART OF ... WARD

St. Anne & Agnes

ALDERSGATE STREET

FIRST PRECINCT

SECOND PRECINCT

THIRD PRECINCT

ST JOHN ZACHARY'S

ST ANNE'S PREC.

SAINT MARTIN'S PRECINCT

ST MARY STAINING PRECINCT

ST LEO'S

ST MARTIN'S LE GRAND

NOBLE STREET

FOSTER LANE

LITTLE BRITAIN

Bartholomew Close

Bull and Mouth Street

Town Ditch

Staining Lane

Maiden Lane

Gutter Lane

Newgate Str.

Blowbladder Str.

Aldersgate

TO The Worshipful Richard Levett Esq. Alderman of this WARD This Plan is Inscrib'd by Your Worship's Most Obedient and most Humble Servant Jacob Ilive.

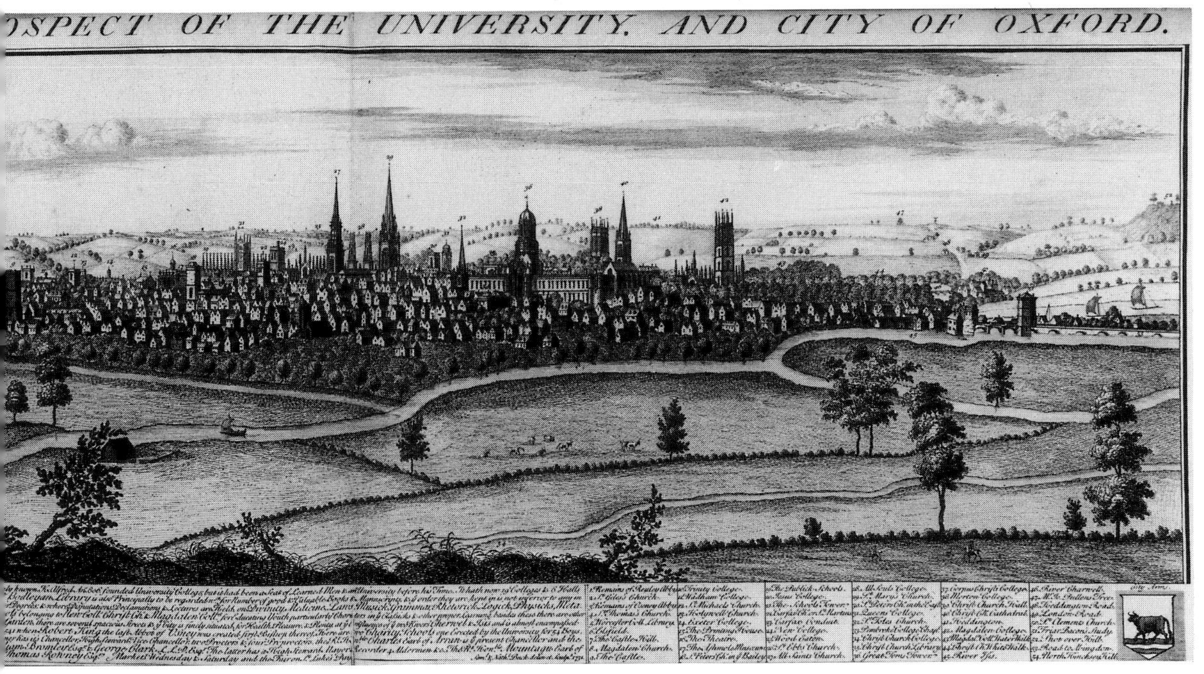

(34) John Rocque, 'The city of Exeter'.
(London: John Rocque, 1744.)
BL Maps K.Top. xi.68.

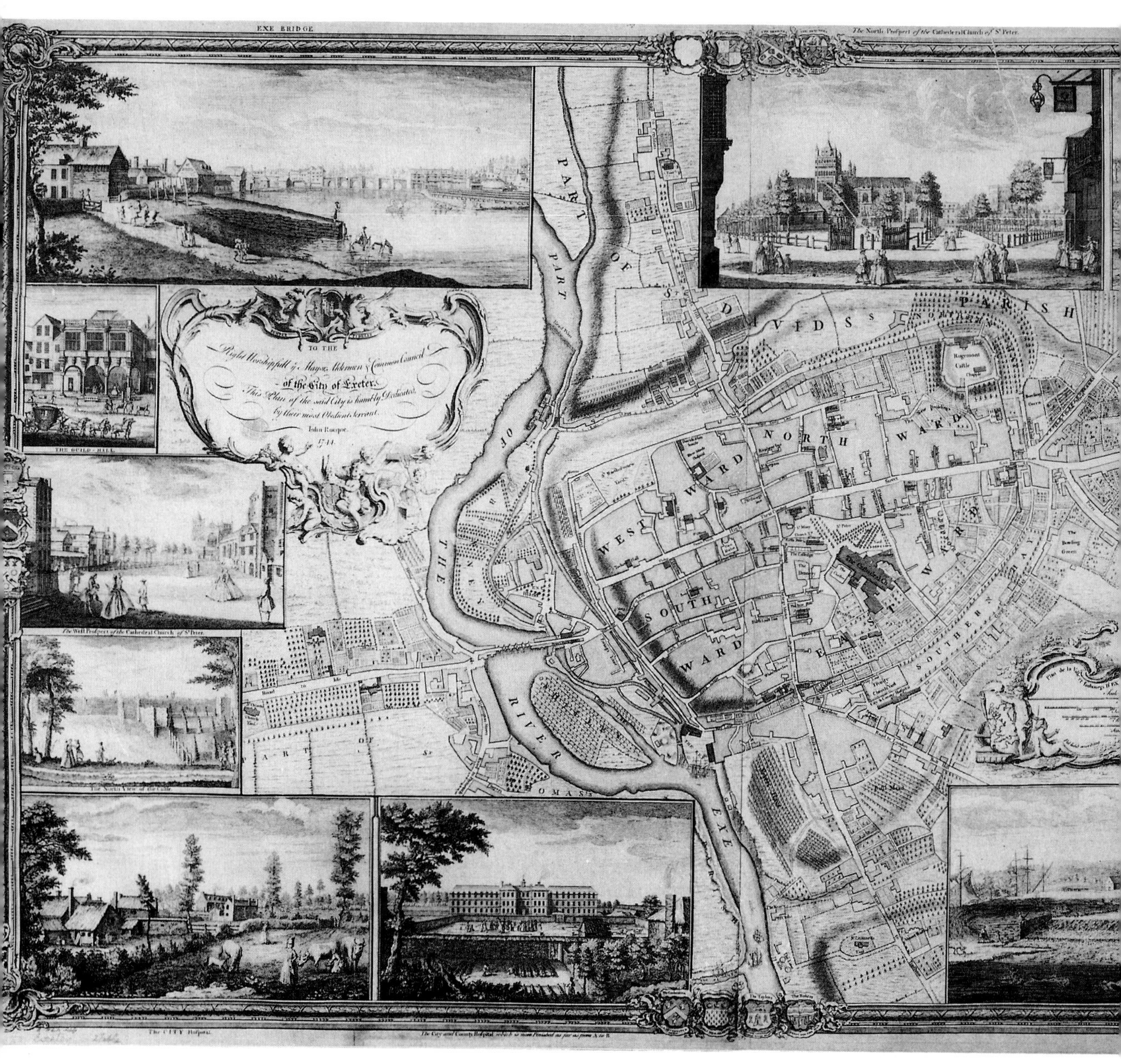

(35) Chichester. From Andrew Dury, *A Collection of Plans of the principal Cities of Great Britain and Ireland.* (London: Andrew Dury, 1766.) BL Maps C.24.a.26.
The Roman origins of Chichester are betrayed by the line of its walls and its cruciform street plan.

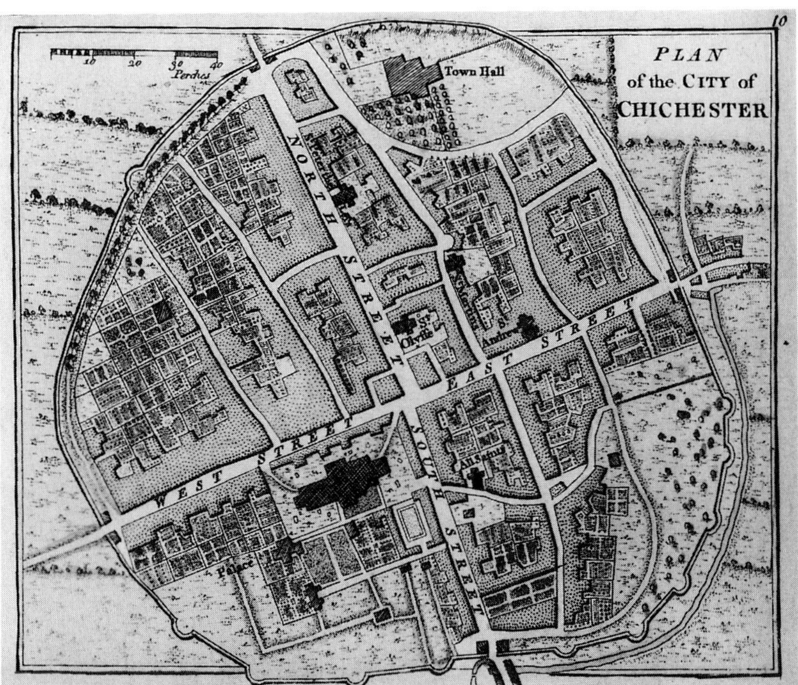

(36) Benjamin Donne, 'A new and correct plan of the City of Bath'. (Bath: H Godwin, 1810.)
BL Maps K.Top XXXVII.23.

5

Industry and Empire: the town plans of the nineteenth century

The growing professional skills of both surveyors and cartographers had developed by the turn of the nineteenth century to such an extent that town and city plans were already utilitarian rather than decorative objects before the process of map production became mechanised. As a reaction to the growing trend towards utility, Thomas Hornor, landscape gardener, surveyor and artist, sought to combine scientific precision with pictorial flair. In his book *Pictorial surveying* he complained, 'the utmost that a common plan can effect . . . is to furnish an outline which the memory must fill up'. His remedy was to depict the subject 'so that the beholder, placed as it were in the air, may enjoy a prospect of which the irregularity of the ground had concealed, perhaps, many pleasing parts'.

Hornor's plan of Kingston-upon-Thames of 1813 (PLATE 9) embodied his philosophy, and was also one of the few maps to employ the aquatint process. Aquatinting consisted of fusing a ground of minute particles of resin to a copper plate to act as a resistant to the acid in which the plate was then immersed. As the resin ground was porous, the acid would bite into the plate in tiny pools around each particle. When the plate was cleared of resin and acid, the minute depressions retained the ink, to give a soft grain which produced a film of shade and tonal variation over the plate. Aquatinting was more commonly associated with topographic prints than with cartography, although Hornor's style, which incorporated an elaborate cartouche as well as views of Kingston and its environs, undoubtedly benefited from the technique.

In the early part of the nineteenth century, most town plans continued to be engraved on copper-plates, either as individual sheets or as collections for atlases such as the *British Atlas* (1810) by J Cole and J Roper, and the two-volume *Maps of the Society for the Diffusion of Useful Knowledge*, published in 1833 and incorporating highly detailed plans of Birmingham, Edinburgh, Liverpool and London as well as major cities elsewhere in the world. Undoubtedly one of the most original urban cartographers operating in Britain in this period was John Wood, a native of Yorkshire who by 1811 had established himself in Edinburgh as a surveyor. Few plans of Scottish towns, apart from the major cities, had appeared before the turn of the century, but Wood rectified this with the publication, between 1819 and 1826, of accurate large-scale plans of over 50 of the smaller communities in Scotland. These were issued in 1828 as the *Town Atlas* (37), along with a companion volume of descriptive text. From 1826 until his death in 1847, Wood surveyed a number of small towns south of the border, notably in Wales and the West Country.

By mid-century, however, urban cartography was being transformed

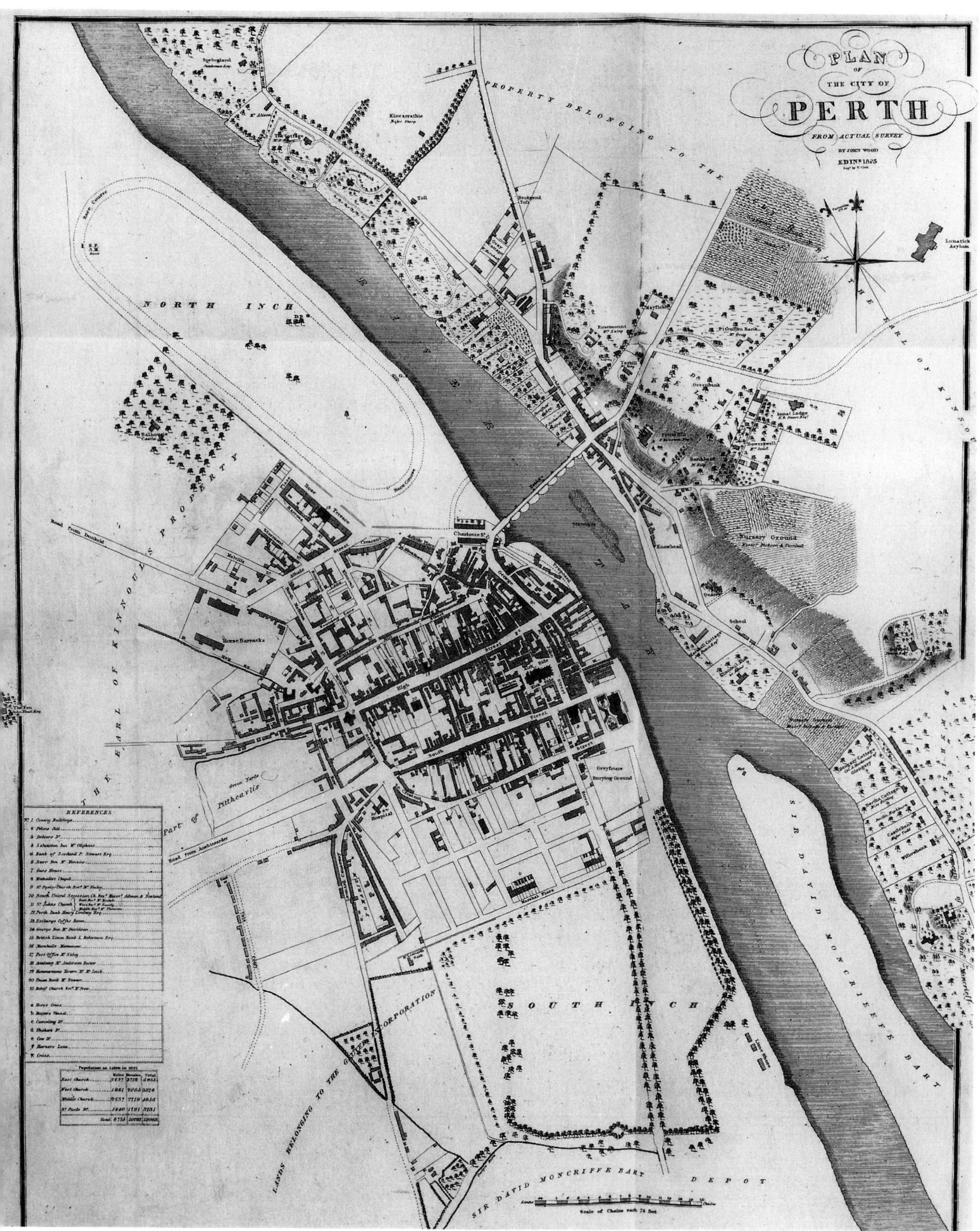
PLAN
OF
THE CITY OF
PERTH
FROM ACTUAL SURVEY
BY JOHN WOOD
EDINR 1823
Lunatick Asylum
NORTH INCH
SOUTH INCH
Charlotte St.
Horse Barracks
Nursery Ground
Greyfriars Burying Ground
Part of Pitheavlis
Road from Dunkeld
Road from Auchterarder
Kinnoull
Bridgend Toll
Toll
Knowhead
School
PROPERTY BELONGING TO THE EARL OF KINNOUL
THE EARL OF KINNOULS PROPERTY
LANDS BELONGING TO THE GUILDRY CORPORATION
SIR DAVID MONCRIEFF BART
SIR DAVID MONCRIFFE BART
DEPOT
REFERENCES
Scale of Chains each 74 feet

(37) John Wood, 'Plan of the city of Perth'. From the *Town Atlas*. (Edinburgh: John Wood & W Swinton, 1823.)
BL Maps C.21.e.4.

by the spectacular urban expansion of the Industrial Revolution. In north-west Europe and North America the growing cities provided a ready market for cheap, up-to-date street plans, which in the new acquistive age were as unadorned as utility permitted. Rapid urban growth also meant that town plans would quickly become out of date, thus presenting a new requirement on the cartographer to revise and reissue his plans to keep pace with the demands of a more mobile and literate urban population. The introduction of lithography from the 1820s onwards provided part of the answer to this demand, as it made the process of map-making cheaper and easier than the highly skilled, laborious and costly method of engraving on copper. Lithography, pioneered in 1796 by Alois Senefelder of Munich, was a chemical process which exploited the ability of porous stone to attract both grease and water, which had a chemical antipathy to one another. The technique required that a design be transferred to a polished stone using a greasy drawing medium. The stone was then washed with water which was absorbed only by those areas of the stone not covered with the medium. A greasy printing ink was then applied by roller, which adhered only to the drawing medium. Prints were obtained by placing the paper face down on the stone and passing stone and paper together through a flat-bed scraper press which would rub the back of the paper as it passed through. Although the technique was widely used by the 1820s, its potential for mass production only became fully realised from the 1870s onwards, when steam-powered mechanical presses were introduced. These picked up the design from the stone and transferred it to the paper by means of rubber rollers, a process known as offset lithography.

The new techniques were extensively exploited in the United States and Canada, where the large-scale mapping of urban America took the form of commercially-produced county and city atlases. These were usually produced on a subscription basis, in which an agent of the publishing company would endeavour to raise money for the project from the local citizenry. This done, field surveys were carried out and supplemented with ownership and property details from tax lists, cadastral maps and other official records. The *Atlas of Buffalo*, 1872, was a typical example and one of the earliest works by G M Hopkins, a company which concentrated its mapping activities in New York, New Jersey, Ohio and Pennsylvania. The firm's base, Philadelphia, was the most important atlas-publishing centre after Chicago.

North America also witnessed a revival of the bird's-eye view on a scale sufficiently widespread to include even the smallest communities. The technique had been used in preceding centuries in Europe to portray the great cities and capitals. In the United States and Canada, however, urban views were mass-produced, not only for the major metropolitan centres, but for over 2,500 lesser towns throughout the continent. That of Winnipeg (**38**) was a typical example of the views that were published between the 1830s and 1920s, satisfying a vigorous demand for domestic wall hangings and serving as publicity material for real-estate agents and chambers of commerce. For each view, the artist would draw from a street plan a projection or framework showing the town in perspective. He would then perambulate each street to make sketches of the

(**38**) 'Bird's-eye view of Winnipeg'. (Winnipeg? W J Bulman, 1900.) BL Maps 70796.(17).
Winnipeg in 1871 was a fur-trading post on the Red River with just 246 inhabitants. Thirty years later, when this view was published, it was a flourishing communications and agricultural centre with a population of over 42,000.

(PLATE 10) *Plan of Huddersfield.* (Southampton: Ordnance Survey, 1890.) BL Maps O.S.T.199
The hand colouring, carried out by apprentice draughtsmen, shows brick and stone buildings in red, iron and wooden buildings in grey, glass covered areas in blue, and streets in buff. Trees, tramlines, walls and fences were all surveyed in their exact position.

The source of Huddersfield's prosperity is indicated by the oval-shaped Cloth Exchange, built in 1769 and dismantled in 1930.

(PLATE 11) Maps of London poverty. From Charles Booth, *Labour and life of the people*, appendix to volume II. (London: Williams & Norgate, 1891.) BL Maps C.21.a.18.

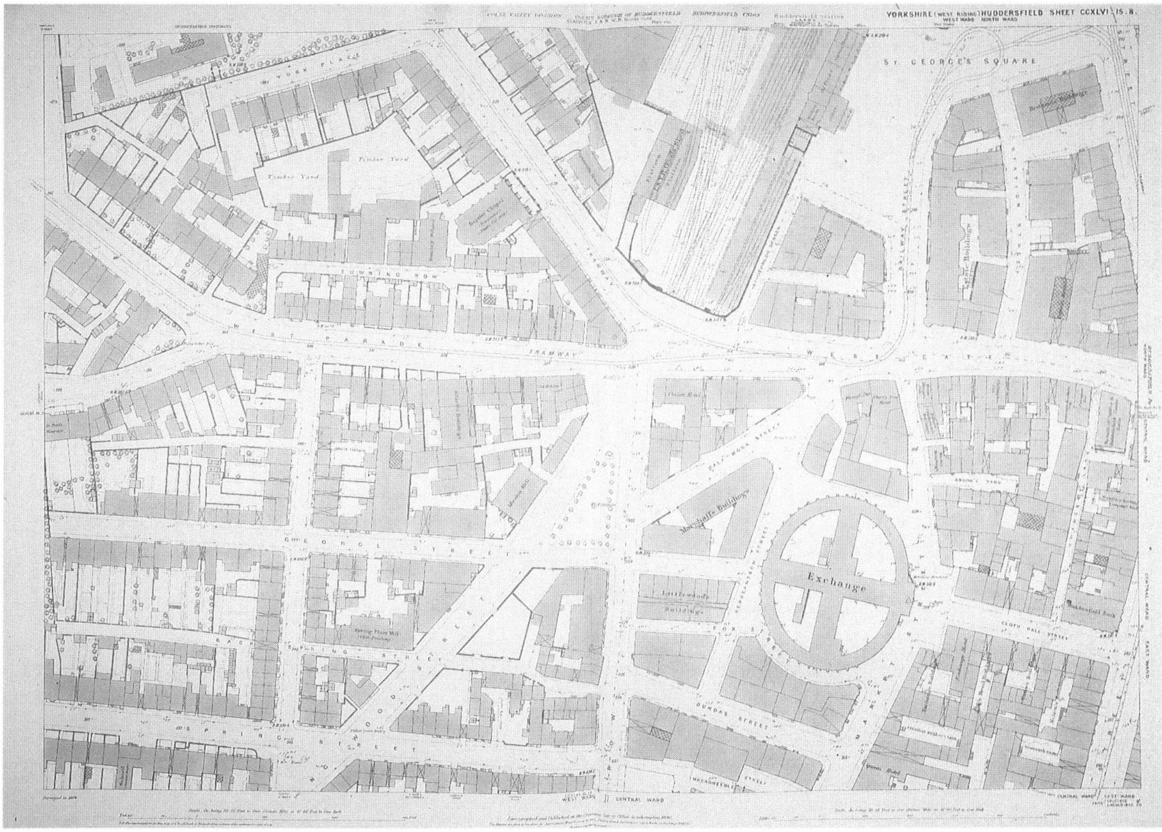

YORKSHIRE (WEST RIDING) HUDDERSFIELD SHEET CCXLVI. 15. 8.

WEST WARD NORTH WARD

ST GEORGE'S SQUARE

YORK PLACE

Timber Yard

TOWNING ROW

WEST PARADE

TRAMWAY

Exchange

Littlewoods Buildings

GEORGE STREET

CLOTH HALL STREET

DUNDAS STREET

SPRING STREET

CENTRAL WARD

EAST WARD

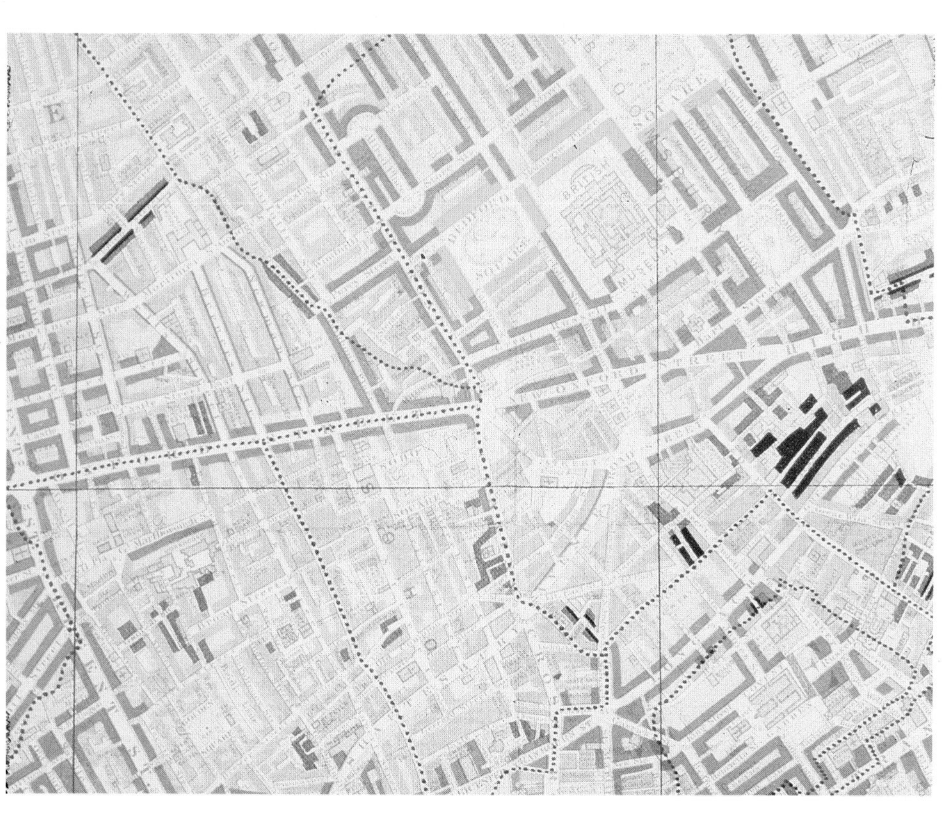

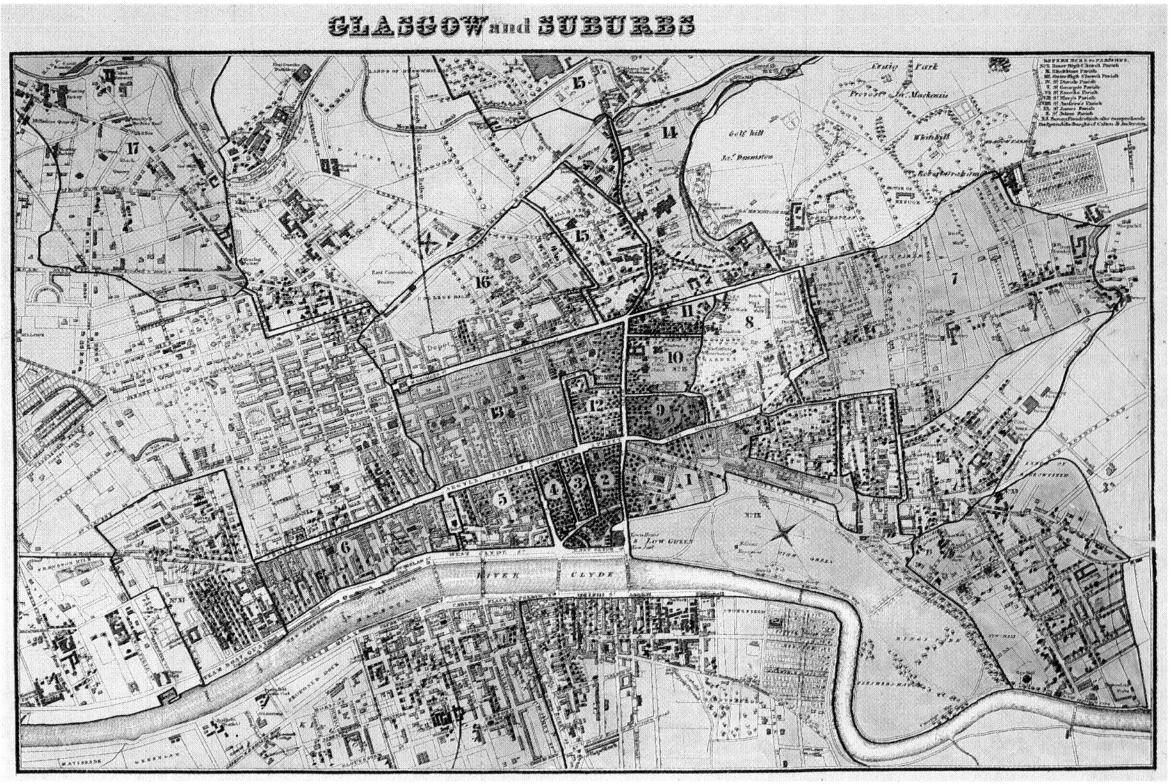

GLASGOW and SUBURBS

(PLATE 12) Map of an epidemic in Glasgow. From Robert Perry, *Facts and observations on the sanitory state of Glasgow*. (Glasgow: Royal Asylum for Lunatics, 1844).
BL C.T.249.18.

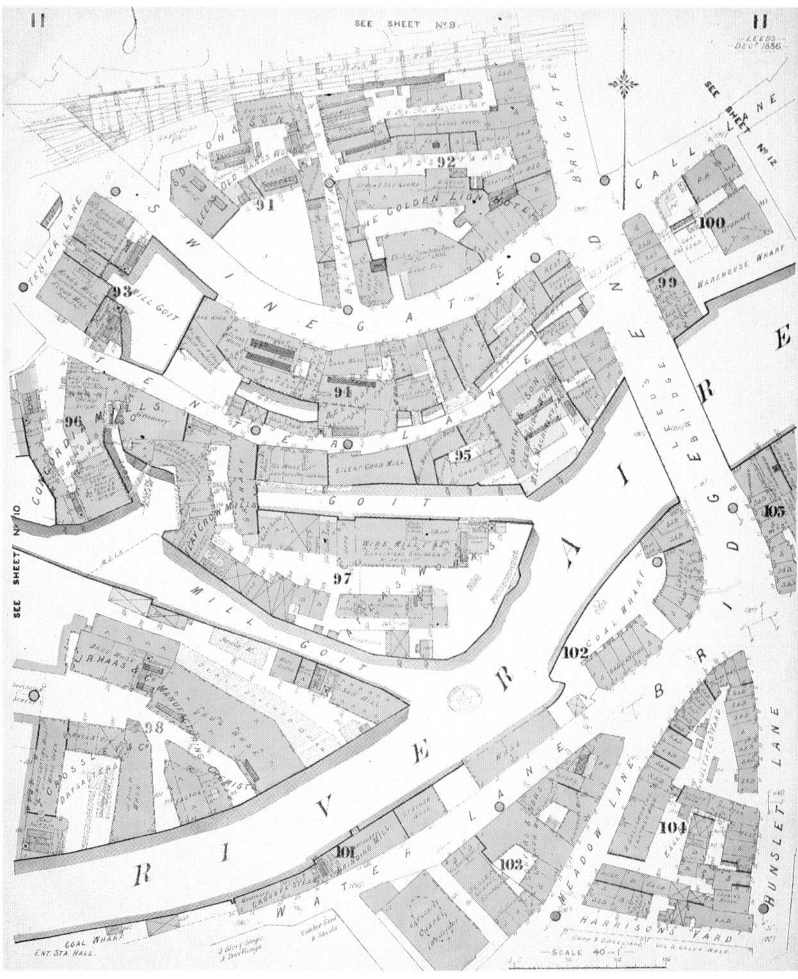

(PLATE 13) *Insurance plan of Leeds*. (London: Chas E Goad, 1886.)
BL Maps 145.b.14.
The plan, drawn to a scale of 60 feet to the inch, shows brick and stone buildings in red, iron and wooden buildings in yellow, and glass sky-lights in blue. Amongst the wealth of detail given are road widths, the location of fire hydrants, the number of storeys of each building, the presence of boilers, windows and doors, the structure of party walls, and the occupant and use of each building.

buildings, which were then redrawn on the projection to show the town as viewed from a high-angle oblique position some 2000–3000 feet above the ground.

In Europe, by contrast, large-scale topographical surveys of urban areas were carried out by official organisations. In Britain, growing anxiety over matters of public health in the major industrial cities led to the establishment, in 1834, of the Poor Law Commission. This body quickly concluded that adequate improvements in urban areas could only be made if suitable maps existed. In response to this need, the Ordnance Survey, then mapping the northern counties of England at the 6 inch scale, was authorised by the Treasury in 1840 to produce a series of town plans at a scale of five feet to one mile, (1:1,056), for towns in the region with a population of 4,000 or more. The first of these plans, covering St Helens, appeared in 1843. A second series, at a scale of 10 feet to one mile (1:528), was published in the early 1850s for a further eighteen towns in England and Wales. These were based on surveys stimulated by the evidence of witnesses to the *First Report of the Commissioners for Inquiring into the State of Large Towns and Populous Districts* (1844) and the Public Health Act of 1848. In 1855, a third series, at a scale of 1:500 (10.56 feet to one mile), was authorised which would cover all towns with a population of more than 4,000. By the end of the programme in 1892, urban Britain had been mapped at scales 'sufficiently large to show detail down to the size of a door-step' (Ordnance Survey, *Annual Report*, 1891). The plan of Huddersfield (PLATE 10), surveyed in 1889 and published the following year on 49 sheets, is typical of the third series plans, and was one of the last to be published before the surveyors moved on to the Irish towns. The availability of accurate large-scale mapping by the Ordnance Survey from the second half of the nineteenth century aided the commercial map producers, who no longer needed to carry out their own surveys. Instead they could produce maps which, with or without acknowledgement, were based on existing Ordnance Survey data. The firm of John Heywood was typical of the age, a commercial publisher whose few maps supplemented a much wider range of printed products from school books, newspapers and technical journals to office stationery and greetings cards. Heywood's 'Pictorial map of Manchester and Salford' of 1886 (**39**) features a ground plan derived from the Ordnance Survey on which has been superimposed elevations of the city's prominent buildings, including Heywood's warehouses and showrooms in Deansgate and his 'Excelsior' printing works.

Among Heywood's contemporaries were specialist cartographic publishers who made their own contribution to urban cartography and whose names remain familiar today. George Washington Bacon was described in the *London Directory*, 1863, as 'the American map publisher'. His company produced a wide variety of maps, atlases and globes, as well as wall charts, lettering charts and other educational aids. He also marketed a 'parlour gymnasium' and trapeze bar, and recruited apprentices to his firm by approaching the London School Board with a prize of £20 and the offer of employment to senior pupils who excelled at map-drawing. His index plan of Margate, like the others of the series, incorporated commercial sponsorship as a means of offsetting the cost of publication.

Another geographical publishing business, George Philip & Son, was founded in Liverpool in 1834 as a bookseller. The firm rapidly moved into publishing and printing, opening offices in London (1848) and new works in Liverpool (1859), equipped with powered lithographic machines. After 1870, when Board Schools were established to provide a national system of education, Philips became one of the leading educational publishers and suppliers, producing a variety of school atlases, charts, ready-reckoners and wall maps. More specialised cartographic works were also produced, such as the plan of Greater Liverpool of 1895.

One of the finest of nineteenth-century street plans was the 12-sheet plan of Edinburgh of 1891 drawn to a scale of 15 inches to one mile (**40**). It was made by John George Bartholomew, who had taken over the family map-publishing business from his father three years previously. The plan was based on the Ordnance Survey's 1876–77 revision of the 1:1056 (five foot) scale town plan, and was produced on new gas-powered mechanical lithographic presses. Sold at 30 shillings, mounted copies were available at 45 shillings on which purchasers could have their home or business identified and coloured at no extra cost.

The nineteenth century also bore witness to the cartographic activities of the official agencies of the colonial powers, which produced plans of many of the major cities as part of their regular activities. In India, for example, the Boards of Revenue were responsible for raising the land taxes necessary to support the British administration. The Sudder Board of Revenue published its maps from standard manuscript atlases which were maintained in all district record rooms at a scale of 4 inches to one mile. A map of Allahabad of 1862, however, owed its origins to military as well as revenue considerations, as the revenue surveyor, Lieutenant-Colonel Strachey, was responsible in 1858 for a report on the position of new civil stations and military cantonments. The report, and map, arose from the need to reorganise both civil administration and military deployment in the wake of the Indian Mutiny.

Despite the near-monopoly of Europe and North America in the production of town plans in the nineteenth century, local traditions of urban mapping continued in both Japan and China throughout the century. This is evidenced by a Chinese woodcut plan of Peking (**41**) of 1900, which nevertheless appears to be a close copy of one made in 1729 to accompany *A description of Peking* by Father Hyacinth Bitchurin of the Russian Ecclesiastical Mission.

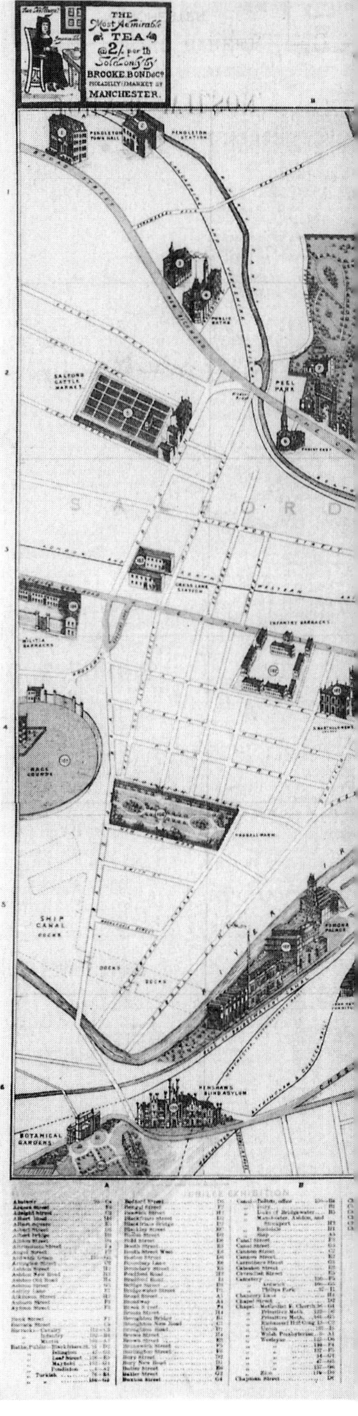

(**39**) 'Pictorial map of Manchester and Salford'. (Manchester: John Heywood, 1886.)
BL Maps 3215.(16).

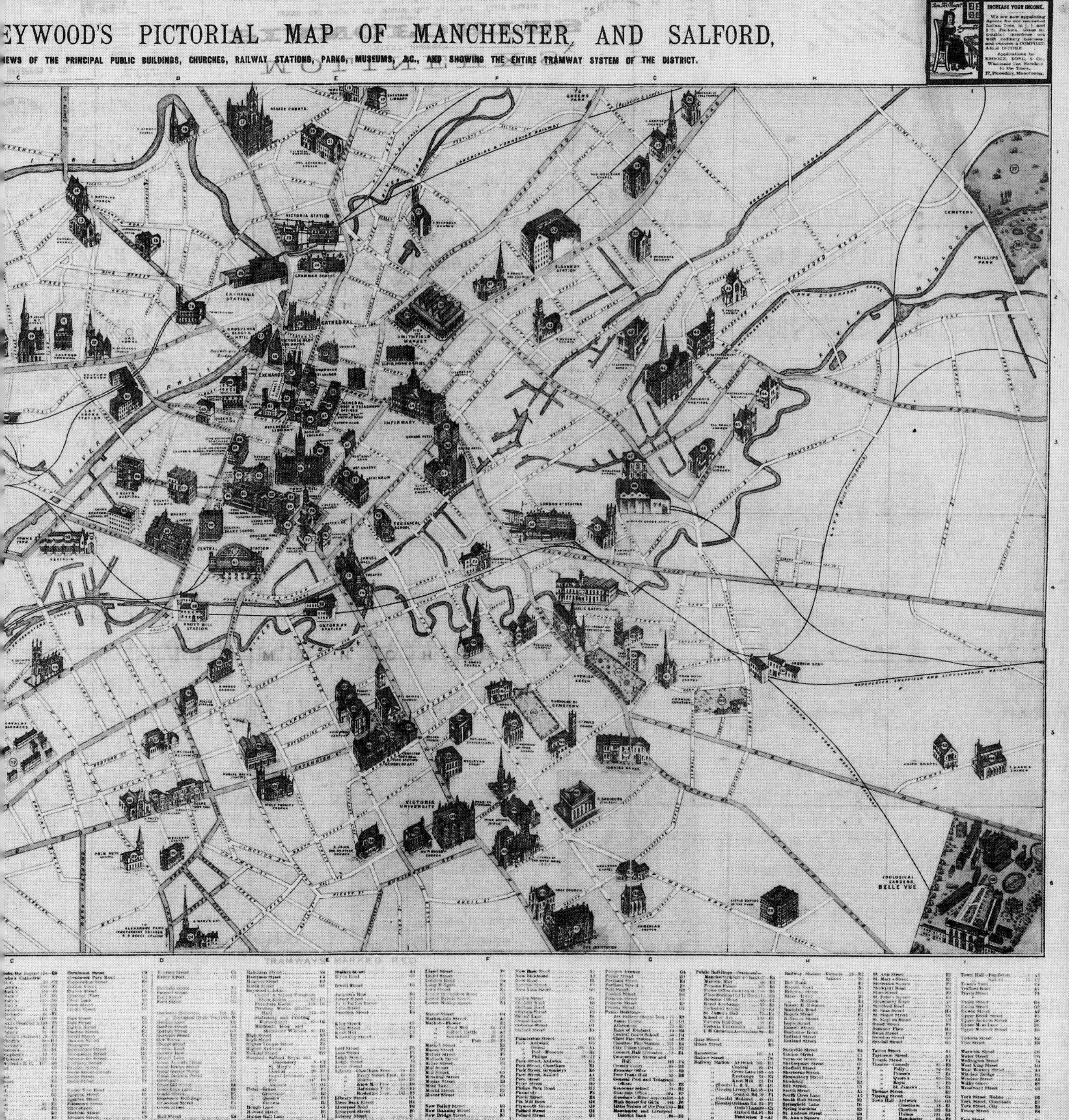

The map shows the major buildings of the city in elevation, and includes the proposed docks of the Manchester Ship Canal (1887–92). The exhibition buildings for Queen Victoria's Jubilee, 1887, have evidently been added (bottom left, corner) in some haste.

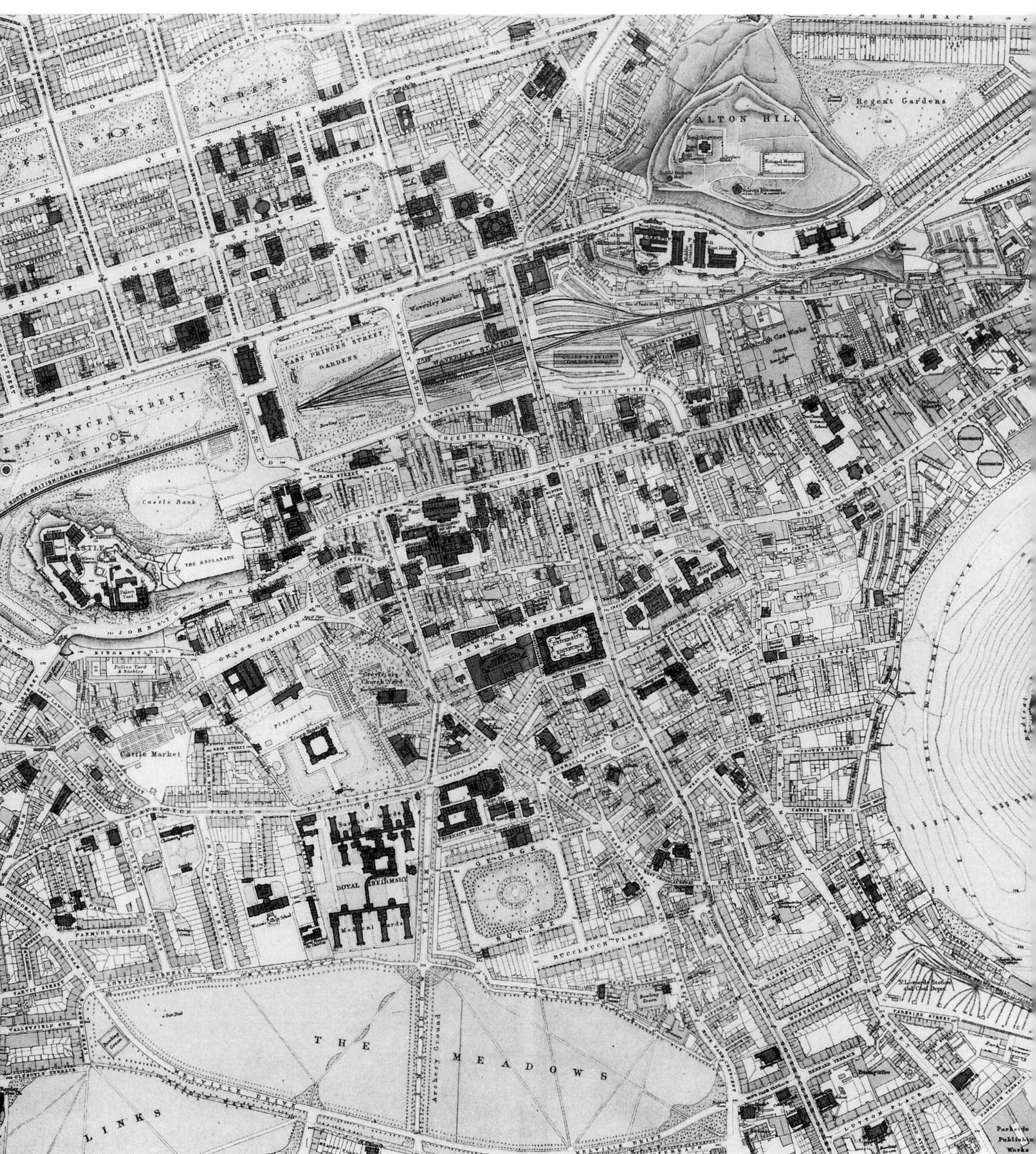

(**40**) John Bartholomew, 'Plan of the city of Edinburgh with Leith and suburbs'. (Edinburgh: John Bartholomew & Co [The Edinburgh Geographical Institute], 1891.)
BL Maps 19.e.16.

(**41**) Anonymous woodcut map of Peking. (Peking? *c.*1900.)
BL Maps 30.b.54.

京師全圖

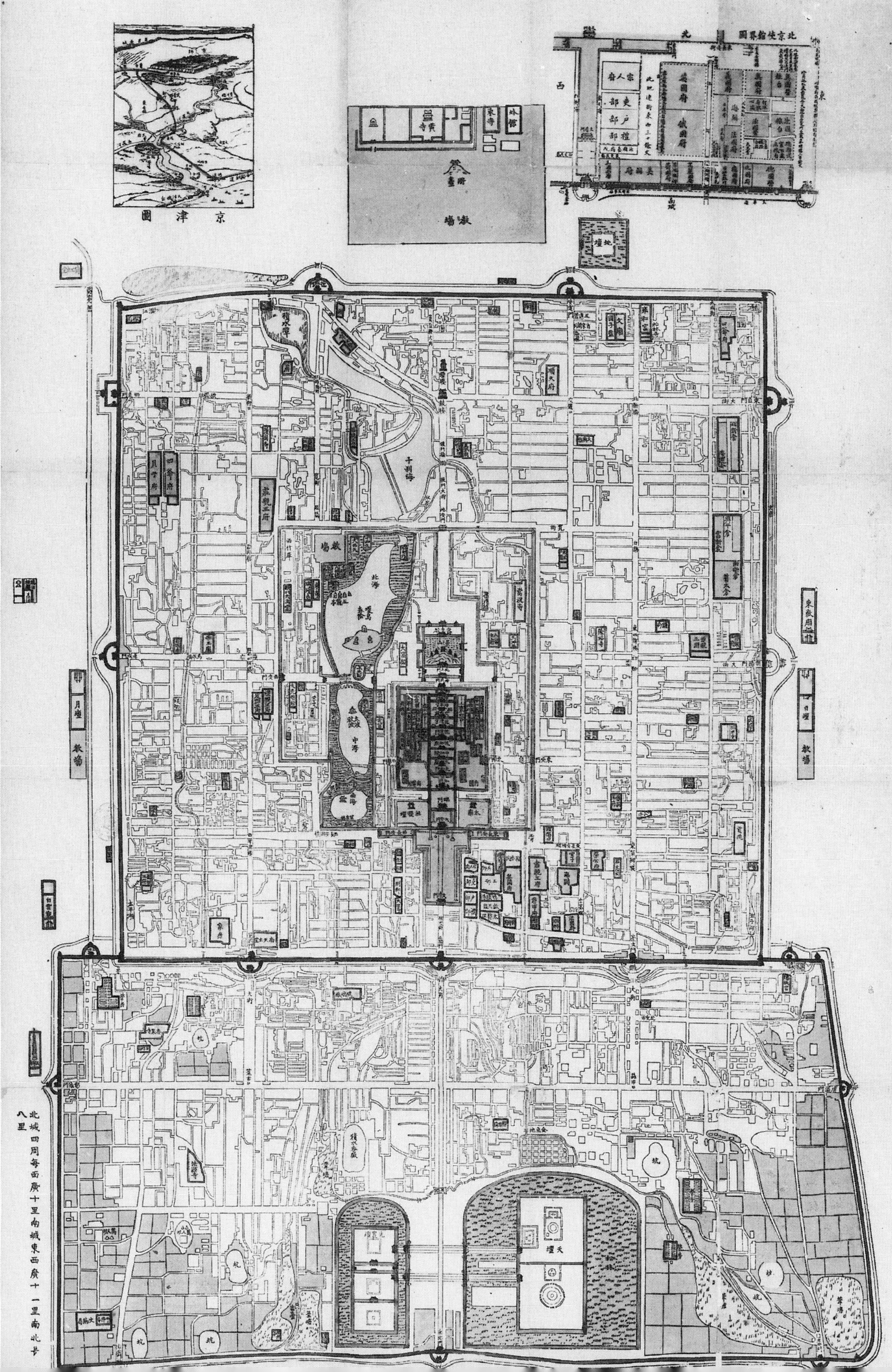
京津圖
北京使館界圖
教場
地壇
北海
中海
南海
什刹海
天壇
先農壇
日壇
月壇
北城四周每面廣十里南城東西廣十一里南北長八里

6

Medical and social mapping

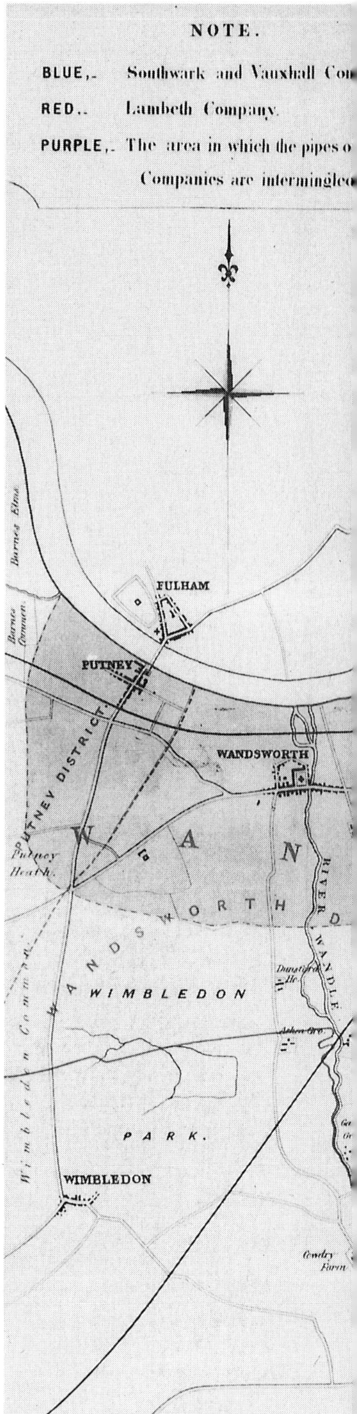

Towns and cities were a fruitful environment for the development of thematic mapping, in which cartography is applied to analyse and explain the underlying social dynamics. In Victorian times, a wide range of issues were explored by maps depicting such phenomena as crime, drunkenness and religious allegiance in major cities. By far the most comprehensive and detailed cartographic analysis of the social environment was, however, that carried out by Charles Booth as part of his monumental study of London poverty (PLATE 11).

Booth, a wealthy London shipowner with Liberal family connections, became interested in the problems of poverty and unemployment in the 1880s. At this time, public interest in these matters was being further stimulated by a growing militancy amongst the working classes and by an awareness developed through the availability of an increasingly large and alarming body of literature which treated the subject through factual reportage and personal anecdote. By 1885, Booth had befriended a group of philanthropists, but was nevertheless sceptical about the real extent of poverty in the capital, and regarded much of the available literature as exaggeration coloured by political and religious bias. He was moved to act when in 1885 the *Pall Mall Gazette* serialised the results of a survey of London's working-class districts which purported to show that some 25 per cent of the capital's population lived in poverty. Early in 1886, Booth resolved to prove objectively that the authors of the survey, the Marxist-orientated Social Democratic Federation, had overstated their case and that their findings were mere propaganda.

In the spring of 1886, he recruited a team of volunteers, including the socialist Beatrice Webb, who collected and analysed data from the police, Poor Law statistics and the reports of school visitors before examining and reporting on every street in the capital. The result was the publication in 1889 of *Life and Labour of the people*, which covered the East End, showing that 35 per cent, not 25 per cent, of the population in this area lived in poverty. The second volume of Booth's survey, entitled *Labour and Life of the People*, covered the rest of London and appeared in 1891. It revealed to Booth's distress that London's poor accounted for no less than 30.7 per cent of its total population. The 1891 volume incorporated a map of the entire city in which the data gathered by Booth's assistants were plotted onto the Ordnance Survey's 25 inch to a mile map of the city, with seven colours to show categories of socio-economic status. These were then sent to the publisher Edward Stanford, who overprinted the colours on his own six-inch 'Stanford's library map of London and its suburbs'.

(42) Map showing the areas served by two London water companies. From John Snow, *On the mode of communication of Cholera*, 2nd edition. (London: Churchill, 1855.)
BL 7560.e.67.

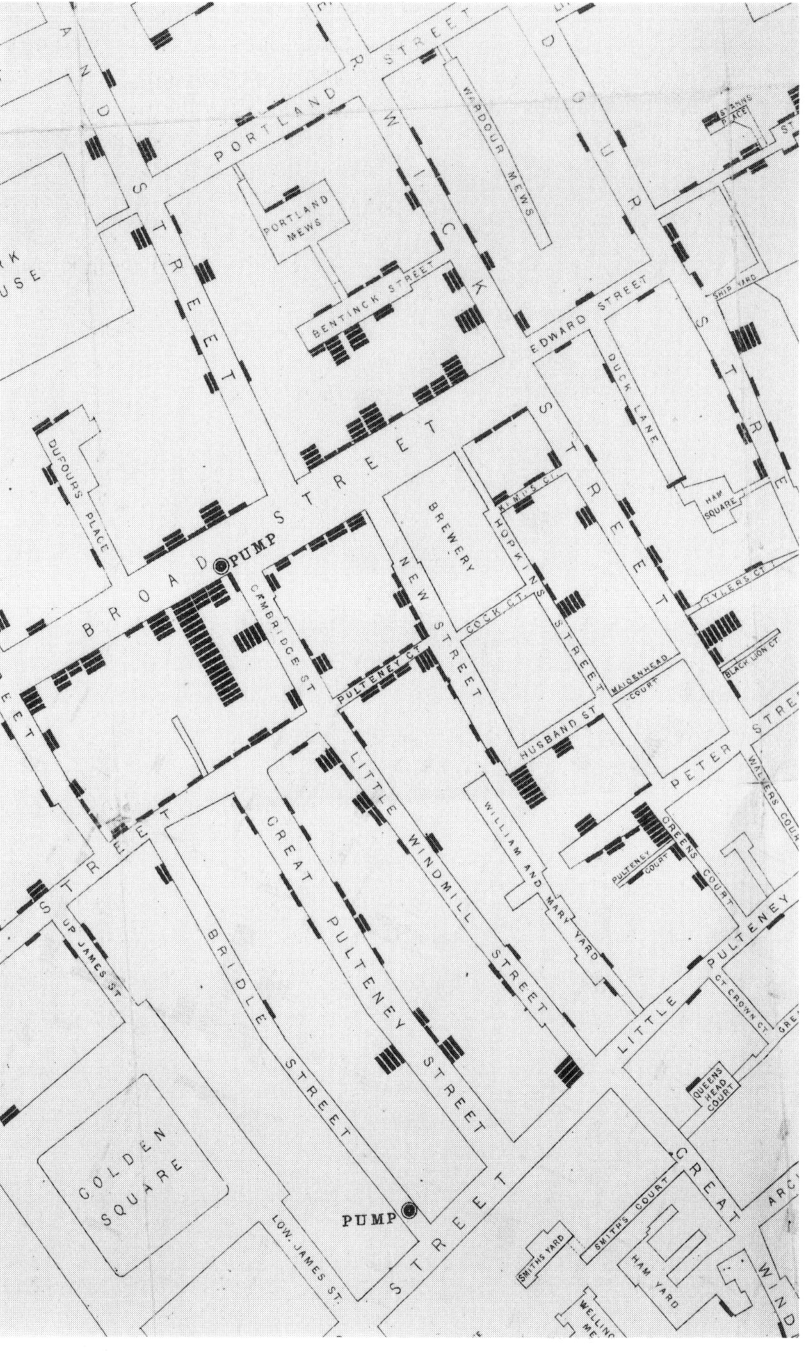

(43) Map showing the distribution of cholera cases in the locality of the Broad Street pump, Soho. From John Snow, *On the mode of communication of Cholera*, 2nd edition. (London: Churchill, 1855.)
BL 7560.e.67.

A specific theme pursued in the urban cartography of the nineteenth century was the distribution and spread of disease. Medical mapping of the urban environment first evolved as the 'spot maps' of North American cities. There, the etiology of yellow fever became of interest to the medical profession in the period 1790–1820 as manifested in studies of the cities of New York, Natchez, New Orleans and others. Nicolas Chervin was the leading French advocate of the 'anti-contagionist' view that yellow fever was not infectious but was transmitted by some carrier.

He became familiar with the use of cartographic representations of the disease during the debates to which he contributed in the 1820s, and was appointed to the commission supported by both French and British governments to study the yellow fever epidemic of 1828 in Gibraltar. Chervin's survey techniques, incorporating interviews with the occupants of every household in the colony, were of unprecedented thoroughness. The result was the two-volume *Documens relatifs à l'épidémie de fièvre jaune*, 1830, whose four plans include one of Catalan Bay showing the locations of sewers and privies in relation to households affected by the disease.

In western Europe the devastating cholera epidemic of 1832 was a major stimulus to the use of maps in the medical reports of city health officials, a procedure continued in subsequent outbreaks in the 1840s and 1850s. The techniques used to portray conditions of public health in urban areas became more sophisticated as the number of published reports increased. A map (PLATE 12) accompanied a tract of 1844 by Robert Perry, Senior Physician to the Glasgow Royal Infirmary, in which the course of an influenza epidemic is traced, using reports collected from the district surgeons of the city: 'I have laid down and numbered the different districts upon a map of the city . . .', he wrote, 'marking with a darker shade those parts where the epidemic was most particularly prevalent, which shows that those places most densely populated, by the poorest people, have suffered most severely'. The report and map served another purpose; an accompanying slip bears the note: 'In order to aid the laudable design of Dr Hutchinson in exercising the mental and bodily faculties of the inmates of the Lunatic Asylum, The Printing of this Paper, the Colouring of Maps, &c., is wholly the work of the inmates'.

Two of the most influential urban medical maps were those in Dr John Snow's essay *On the mode of communication of cholera*, 2nd edition, 1855. Snow was convinced that cholera was carried by contaminated water, and his report contained a map (**42**) accompanied by statistical tables setting out the death rates in the areas served, respectively, by the Southwark & Vauxhall Water Company and the Lambeth Water Company. In 1852 the Lambeth Company had changed its source of supply to Thames Ditton, which was free of pollution from London sewage. In the red-shaded areas served exclusively by the Lambeth Company, there were only five deaths per 1000 households, while the blue-shaded Southwark & Vauxhall Company's areas, served by water from the Thames at London Bridge, had a mortality rate of 71 deaths per 1000.

Snow (himself a teetotaller) is remembered today in the name of the John Snow public house in Soho, on the site of the original Broad Street pump, which he was able to pinpoint as a source of cholera by means of the report's second map (**43**). This showed the location of the pumps in the Soho area and the distribution of cholera cases, the pattern of which revealed a great concentration around the Broad Street pump.

7

'Fire and the sword'

The medical and social mapping of urban centres was primarily a nineteenth-century phenomenon which flourished in an age when the problems of large cities and industrialised societies were of such a scale that new approaches, and new techniques of investigation, were required to understand them. Maps thus became a tool with which to analyse urban processes. Thematic mapping of urban areas had, however, long predated the nineteenth century in its concern with specific events in a city's history, such as sieges, fires and other disasters.

The frequency of sieges in military campaigns arose from the presence on the battlefield of the fortress and fortified town, which provided a degree of stability at a time when armies were small, ill-regulated and relatively immobile. Siege warfare continued to be a central element in military geography well into the eighteenth century. Most of the major sieges of recent history have been the subject of the map-makers' attention, for the military campaign and the town plan alike were traditional subjects for the cartographers' skills. Siege maps, moreover, could be used both as news and propaganda, as demonstrated by de Hooghe's bird's-eye view of the investment of Luxembourg in 1684 (44). The city had been beseiged after Louis XIV ordered French troops to occupy the Duchy in 1681 as part of his policy to extend the frontiers of France. In order to establish some legal basis for his claims, special courts known as Chambres de Réunion had been set up, whose task was to examine old deeds and charters for evidence that a particular territory had once belonged to France. When the Metz Chambre de Réunion decided in 1681 that Louis had such a claim to the Duchy of Luxembourg, the province was invaded. The city finally capitulated in June 1684 after a desperate siege culminating in a devastating bombardment. Romeyn de Hooghe, an Amsterdam engraver and draughtsman specialising in political and religious subjects as well as views and plans of the great events of his day, produced an etching showing a bird's-eye view of the final stages of the siege.

From England came an anonymous broadsheet featuring a generalised map of the siege of Colchester of 1648. This began when George Goring, Earl of Norwich, was trapped in the town after he and Sir William Waller had unsuccessfully attempted to stage a Royalist rebellion in Kent and Essex. The Parliamentary commander Lord Fairfax invested the town on 14 June, and the Royalists surrendered on 22 August. The author of the map was with the Parliamentary forces, as revealed by his comment that 'the word of the King's forces at ye fight was Charles, the ground they fought upon, Maries: ours, God's – our help'.

(44) Romeyn de Hooghe, 'Luxemburgum'. (Amsterdam: Nicolas Visscher, 1684.)
BL Maps C.9.e.4(63).
The view of Luxembourg shows the city besieged by the French army in 1684. The symbolic cartouche, bottom centre, shows figures to the right representing France (holding a shield with a portrayal of the sun king, Louis XIV) and war banishing the figure of justice, bearing a pair of scales, to the left.

Abundant evidence of political bias was evident too in Abel Boyer's plan of Preston. In October 1715, a small force of Scottish Highlanders and English Jacobites assembled in Northumberland as part of the general uprising in support of the claim to the throne of James Edward Stuart, the 'Old Pretender'. The rebels crossed into Lancashire, hoping for assistance from both Roman Catholics and High Anglican Tories. In the event the support failed to materialise, and the incompetent leadership of Thomas Foster led to an ignominious surrender at Preston on 14 November. Boyer, a French Huguenot who had arrived in England in 1689, was a journalist, theatrical producer and compiler of one of the best English-French dictionaries of his day. He also published *The Political State*, a monthly journal giving, for the first time, regular details of the debates in the Houses of Parliament. His zealous Whig inclinations and support of the Hanoverian succession explained his satisfaction at the defeat of the rebels; the cartouche of his map was decorated with bound and bedraggled Scots and Jacobites.

A later and more elaborate military depiction was 'Paris en Mai 1871', published to show the plan of campaign of the French government army in the suppression of the Paris Commune. It reveals how much both urban planning and military considerations were taken into account by Baron Haussman in 1859, who, as prefect of the Seine, directed the rebuilding of the city under Napoleon III. Haussman's design incorporated wide boulevards which were difficult to barricade. At the same time, they offered excellent fields of fire for artillery, while allowing large bodies of troops and cavalry to move easily along them. The flanks of any obstacle could be turned by means of obliquely-intersecting streets and the vast star-shaped *places* with their radiating routes. It is significant that the government troops first entered the rebuilt western areas of the city, and that the last stand of the Communards was in the warren of old narrow streets of the 20th Arrondisment around the Pere-Lachaise cemetery.

The devastating effect of large urban fires, before the introduction of building regulations, proper fire services and mains water supply, also stimulated cartographic activity. In the wake of the Great Fire of London, 1666, many surveyors and cartographers were employed to map the destroyed city and the replacement rising from the ashes. The fire which started in a baker's chimney in Pudding Lane on 2 September 1666, and which Samuel Pepys in his diary described as the 'most horrid, malicious and bloody flame', eventually destroyed some 13,200 houses, 87 churches, the halls of 44 livery companies, and major buildings such as the Royal Exchange, the Guildhall and St Paul's Cathedral. The Czech artist and engraver Wenceslaus Hollar, Royal Scenographer to Charles II, had more than a mere professional interest in the Great Fire. He had spent seven years and had run into debt while preparing an enormous plan-view of the city ten feet by five, when the disaster rendered his surveys and drawings obsolete. His etchings of the survey of John Leake and others (**45**) were a response to the widespread public interest aroused in Europe. The map showed the maximum extent of the destruction, with the sites of the major public buildings, including old St Paul's Cathedral, clearly marked. The halls of the city livery companies were denoted by their coats of arms. Ten years after the fire,

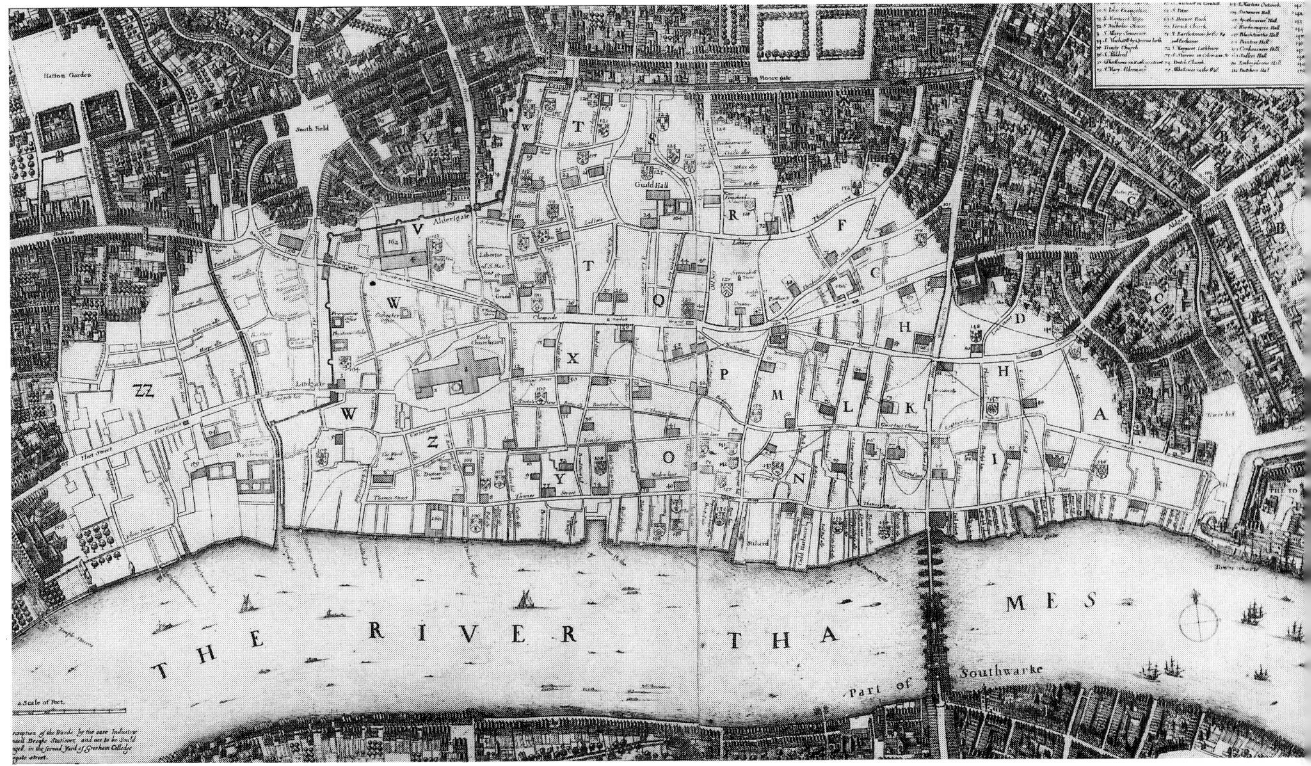

the plan of John Ogilby and William Morgan of 1676 (**26**) showed how quickly the capital had been rebuilt.

The danger of fire prompted, in later years of the nineteenth century, the large-scale production of fire insurance plan by companies such as Charles E Goad in Canada and Britain (PLATE 13) and the Sanborn Map Company in the United States. Fire insurance plans developed from the specific requirements of insurance underwriters, who needed accurate large-scale maps of the major urban centres to enable them to assess the risk attached to an insurance proposal for any given building. Also, in an age where large fires were much more common than today, they had to be aware of the geographical spread of their clients' properties in order to avoid concentrating all their risks in one group of buildings.

Goad, born in London in 1848, moved to Canada in 1869 and published fire insurance plans of all Canadian towns and cities with a population of over 500. He returned to England in 1886. His London company dominated the production of British fire insurance plans, and covered 126 areas including the centres of most major cities by means of coloured large-scale plans which continued to be revised until 1970.

The fire insurance plans represented the last word in specialisation and utility towards which urban cartographers had progressed since the fourth century BC. The simple outlines on the clay tablets of the Babylonians evolved through the symbolic bird's-eye views of the Renaissance into the precise and functional street-finders of the nineteenth and twentieth centuries. The plans of the cities they depict reflect the role of the city as the hub of civilisation. As the poet Paul Claudel writes, 'la ville est la forme de l'humanité'.

(**45**) 'An exact surveigh of the . . . ruines of the City of London'. Wencelaus Hollar *fecit.* (London: John Leake, 1669.)
BL Maps Crace I.50.
The map shows the extent of the devastation caused by the Great Fire, 1666. The sites of the major public buildings, including St Paul's Cathedral, are clearly marked, while the destroyed halls of the city livery companies are denoted by their coats of arms. [See *frontispiece*.]

Further reading

ALLARD, C. *Orbis habitabilis oppida et vestitus* (facsimile). Introduction by R A Skelton. Amsterdam, 1966.

BACHMANN, F. *Die alten Städtebilder*. Stuttgart, 1965.

BRAUN, G and HOGENBERG, F. *Civitates Orbis Terrarum* (facsimile). Introduction by R A Skelton. 3 vols. Amsterdam, 1966.

BRETEZ, L. *Le Plan de Louis Bretez dit 'Plan de Turgot'* (facsimile). Paris, 1979.

CARTER, H. *An Introduction to urban historical geography*. London, 1983.

CLARK, P and SLACK, P. *English towns in transition, 1500–1700*. Oxford, 1976.

CONSTABLE, K M. 'The early printed plans of Exeter, 1587–1724'. *Transactions of the Devonshire Association*, vol.lxiv (1932), pp.455–73

CORTAZZI, H. *Isles of gold: antique maps of Japan*. New York, 1983.

DILKE, O A M. *Greek and Roman maps*. London, 1985.

FORDHAM, A. 'Town plans of the British Isles'. *Map Collectors' Series*, no.22. London, 1965.

FRIED, A and ELMAN, R M (eds). *Charles Booth's London*. London, 1969.

GARDINER, L. *Bartholomew 150 years*. Edinburgh, 1976.

GUTKIND, E A. *International history of city development*. 6 vols. New York, London, 1970–2.

HARLEY, J B. *Maps for the local historian*. London, 1972.

HARVEY, P D A. *The history of topographical maps: symbols, pictures and surveys*. London, 1980.

HODGKISS, A G. *Understanding maps: a systematic history of their use and development*. Folkestone, 1981.

HOWGEGO, J. *Printed maps of London circa 1553–1850*. Folkestone, 1978.

HYDE, R. 'Cartographers versus the Demon Drink'. *Map Collector*, no.3 (1978) pp.22–27.

HYDE, R. 'London ward maps'. *Map Collectors' Series*, no.38. London, 1967.

HYDE, R. *Printed maps of Victorian London 1851–1900*. Folkestone, 1975.

HYDE, R. 'Thomas Hornor: pictural land surveyor'. *Imago Mundi*, vol.xxix (1977), pp.23–34.

JARCHO, S. 'Yellow fever, cholera, and the beginning of medical cartography'. *Journal of the history of medicine and allied sciences*, vol.xxv (1970), pp.131–42.

KOEMAN, C. *Geschiedenis von de kartografie von Nederland*. Alphen aan den Rijn, 1983.

LINKS, J G. *Townscape painting and drawing*. London, 1972.

LOBEL, M D (ed). *The atlas of historic towns*, vol.2 (Bristol, Cambridge, Coventry, Norwich). London, 1975.

MOIR, D G. *The early maps of Scotland to 1850*, vol.2. Edinburgh, 1983.

MUMFORD, L. *The city in history*. London, 1966.

MÜNSTER, S. *Cosmographei* (facsimile). Introduction by Professor Dr R Oehme. Amsterdam, 1968.

OGILBY, J and MORGAN, W. *A large and accurate map of the city of London* (facsimile). Introductory notes by Ralph Hyde. Lympne Castle (Kent), 1976.

PHELPS STOKES, I N. *The iconography of Manhattan Island 1498–1909*. 6 vols. New York, 1915.

PHILLIPS, H. 'John Rocque's career'. *London Topographical Record* vol.xx (1952) pp.9–25.

PRITCHARD, J E. 'Old plans and views of Bristol'. *Transactions of the Bristol and Gloucestershire Archaeology Society*, vol.xlviii (1921) pp.325–53.

REPS, J W. *The making of urban America: a history of city planning in the United States*. Princeton, NJ, 1965.

REPS, J W. *Views and viewmakers of urban America*. Columbia, 1984.

ROBINSON, A H. *Early thematic mapping in the history of cartography*. Chicago, 1982.

ROWLEY, G. *British fire insurance plans*. Hatfield, 1984.

ROWLEY, G. 'An introduction to British fire insurance plans'. *Map Collector*, no.29 (1984), pp.14–21.

SCHULZ, J. 'Jacopo de' Barbari's view of Venice: map making, city views and moralized geography before the year 1500'. *Art Bulletin*, vol.lx (1978), pp.425–74.

SKELTON, R A. 'Tudor town plans in John Speed's *Theatre*'. *Archaeological Journal*, vol.cviii (1952), pp.109–120.

SMITH, D. *Victorian maps of the British Isles*. London, 1985.

STEPHEN, G A. *A descriptive list of Norwich plans 1541–1914*. Norwich, 1928.

TOOLEY, R V. 'North American city plans: a selection', *Map Collectors' Series*, no.20. London, 1965.

UNGER, E. 'Ancient Babylonian maps and plans'. *Antiquity*, vol.ix (1935), pp.311–22.

VARLEY, J. 'John Rocque. Engraver, surveyor, cartographer and map-seller'. *Imago Mundi*, vol.v (1948), pp.83–91.

VAUGHAN, R. *Matthew Paris*. Cambridge, 1958.

WHEATLEY, H B and ASHBEE, E W (eds). *The Particuler Description of England 1588*. London, 1879.

WILSON, A. *The making of the Nuremberg chronicle*. Amsterdam, 1977.

WOODWARD, D. *Five centuries of map printing*. Chicago, 1975.

WRIGHT, N (ed). *Great cities of the world*. London, 1978.

List of exhibits

NOTE: Numbers in **bold** refer to the exhibition; *italic* numbers in brackets refer to the illustrations and plates in the book.

1. Fragment of a city labelled 'Tuba' drawn on clay, 4th century BC (*1.*) British Museum, Department of Western Asiatic Antiquities, no. 35385.
2. Lari, Muhyi: Futuh-al-Haramayn, late 16th or early 17th centuries; MS. (*4.*) BL OMPB Oriental 11533, f.18v.
3. 12th century copy of a Roman text on estate surveying; MS. (*3.*) BL Add MS 47679, ff.63v–64.
4. Vesconte, Pietro: maps of Jerusalem and Acre. *In* Torsello, Sanudo: 'Liber Secretorum fidelium crucis', *c.*1320–25; MS. (*5.*) BL Add. MS 27376*, ff.189v–190.
5. Depiction of Constantinople. *In* Buondelmonte, Cristoforo: 'Liber Insularum Archipelago', 1482; MS. (*Plate 1.*) BL Arundel MS 93, ff.154v–155.
6. Reuwich, Erhard: View of Rhodes. *From* Breydenbach, Bernhard von: *Peregrinatio in Terram Sanctam*, 1486. (*6.*) BL C.20.e.3.
7. Plan-view of Mexico. *From* Cortès, Hernando: *La preclara Narratione dei Ferdinando Cortese della Nuova Hispana*, 1524. (*7.*) BL 9771.b.11.
8. Views of Treves and Metz. *In* Münster, Sebastian: *Cosmographia*, 1544, (*9.*) BL Maps C.8.a14, pp.lx–lxi.
9. View of Solothurn. *In* Stumpf, Johanes: *Die gemeiner loblicher Eydgnoschafft Stetten, Landen und Volckeren Chronik wirdiger thaaten beschreybung*, 1548. (*10.*) BL 1482.g.9. Vol.2, f.223v.
10. Anthoniszoon, Cornelis: 'De vermaerde Koopstadt van Amstelredam', 1544. (*Plate 2*) BL Maps S.T.A.(4).
11. Lafreri, Antonio: 'Le sette chiese di Roma', 1575. (*12.*) BL Maps 23807.(1).
12. Barbari, Jacopo de': 'Venetie', 1500. (*11.*) British Museum, Department of Prints and Drawings, 1895–1–22–1192.
13. Barreto de Resende, Pedro: Plan of Goa. *From* Boccaro, Antonio: 'Livro do Estade da India Oriental', 1646; MS. (*Plate 4*) BL Sloane MS 197, ff.247v–248.
14. 'The Royall City of Tangier in Africa'. *From* Seller, John: *Atlas Maritimus*, 1675. BL Maps C.8.b.13.
15. Merian, Matthäus: 'Civitatis Avenionis'. From an untitled collection of maps and city views, 1635. BL Maps K.7. Tab.33.
16. Plan of Cologne. *From* Braun, Georg: *Civitates Orbis Terrarum*, book 1, 1572 (edition of 1623?) (*Plate 3.*) BL Maps C.29.e.1. (1), f.78.
17. 'Palma'. *From* Braun, Georg: *Civitates Orbis Terrarum*, book 5, 1598 (edition of 1623?). (*13.*) BL Maps C.29.e.1. (3), f.68.
18. Braun, Georg: Letter to Abraham Ortelius, 31 October 1571; MS. BL Harley MS 7011, f.167.
19. Burton, Robert: *The Anatomy of Melancholy*, 1621. BL C.45.c.30, p.351.
20. Lautensack, Hans: 'Warhafftige Contrafactur der löblichen Reychstat Nuremberg', 1552. (*14.*) BL Maps 28840.(7).
21. Anon: 'A Description of the Towne of Mannados or New Amsterdam as it was in September 1661' (the 'Duke's Plan' of New York), 1664. (*Front cover.*) BL Maps K.Top CXXI.35.
22. 'Argentina. Strassburg'. *From* Jansson, Jan: *Theatrum exhibens illustriores principesque Germaniae Superioris civitates*, 1657. (*15.*) BL Maps C.25.b.15. Vol.1,f.6.
23. 'Legia sive Leodium, vulgo, Liege'. *From* Blaeu, Joan: *Toonneel der Steden van's Konings Nederlanden*, 1649. (*16.*) BL Maps C.4.d.3.
24. Paired views of Cochin and its inhabitants. *From* Allard, Carel: *Orbis habitabilis oppida et vestitus*, *c.*1695. (*17, 18.*) BL Maps C.24.e.23.
25. Anon: woodcut map of Nagasaki, *c.*1680. (*19.*) BL OMPB Oriental 75.g.25.
26. Introductory section of the itinerary from London to Apulia. *From* Paris, Matthew: 'Historia Anglorum', after 1252; MS. (*Plate 5.*) BL Royal MS 14.c.VII, f.2.
27. 'Nordovicum Angliae Civitas Anno. 1558 I.B.F.' (Norwich) *From* Cuningham, William: The Cosmographical Glasse, 1559. (*20.*) BL 59.i.28.
28. Lyne, Richard: 'Oppidum Cantebrigiae' (Cambridge). *From* Caius, John: *Historia Cantebrigiensis Academiae*, 1574. BL C.24.a.27.
29. Hooker, John: 'Isca Damnoniorum' (Exeter), 1587. (*Plate 6.*) BL Maps C.5.a.3.
30. Digges, Thomas, or Hill, John: 'The state of Dover Haven with the new workes', 1595; MS. (*24.*) BL Cotton MS Augustus I.i.46.
31. 'Bristow'. *From* Smith, William: 'The Particular Description of England with the portratures of Certaine of the Cheiffest Citties & Townes', 1588; MS. (*21.*) BL Sloane MS 2596, ff.76v–77.
32. Adams, Robert: Plan of Flushing, 1588; MS. BL Cotton MS Augustus I.ii.105.
33. Millerd, James: 'The famous citty of Bristoll', *c.*1710. (*25.*) BL Maps K.Top. XXXVII.32
34. Ogilby, John, and Morgan, William: 'A large and accurate map of London', 1676. (*26.*) BL Maps Crace II.61.
35. 'The West Riding of Yorkeshyre with the most famous and fayre Citie Yorke described'. Speed, John: *From The Theatre of the Empire of Great Britaine*, 1611–12. (*22.*) BL Maps C.7.c.20, ff.79v–80.
36. Norden, John: Plan of the Little Park of Windsor. *From:* 'Description of the Honor of Windesor', 1607; MS. (*23.*) BL Harley MS 3749, ff.5v–6.
37. Briot, Nicholas: cast silver medal commemorating Charles I's return to London from his Scottish Coronation, 1633. Private collection.
38. Hanson, Thomas: 'A plan of Birmingham', 1778. (*31.*) BL Maps K. Top. XLII.81.
39. Baumeister, Johann Jacob: 'Nouveau Plan du Carolsruhe', 1739. (*27.*) BL Maps 30430.(2).
40. Rocque, John: 'The city of Exeter', 1744. (*34.*) BL Maps K.Top. XI.68.
41. Seutter, Matthäus: 'Berlin', 1720. BL Maps 29720.(2).
42. Roxas, D Antonio Fernandez de: 'Topographia de la civdad de Manila', *c.*1739. BL Maps K.Top. CXVI.40.
43. Gordon, Peter: 'A view of the town of Savannah as it stood the 29th March 1734'. (*29.*) BL Maps K.1. Tab.44, f.43.
44. Bretez, Louis: 'Plan de Paris . . . dessiné et gravé sous les ordres de Michel Etienne Turgot', 1739. (*30.*) BL Maps C.11.d.10.
45. Seale, Richard: 'A plan of the ward of Aldersgate', 1739. (*32.*) BL Maps Crace VIII. (2).
46. Buck, Samuel and Nathaniel: 'The south west prospect of the University, and City of Oxford', 1731. (*33.*) BL Maps C.10.a.21.
47. Chichester. *In:* Dury, Andrew: *A Collection of Plans of the principal Cities of Great Britain and Ireland*, 1766. (*35.*) BL Maps C.24.a.26.

48. A Plan of the city of Dublin. *In*: Andrews, John: *A Collection of Plans of the Capital Cities of Europe*, 1771. BL Maps C.26.d.23.
49. Imperial Academy of Sciences: 'Plan de la ville de St. Petersbourg', 1753. (*28*.) BL Maps K.4. Tab.44.
50. Yeakell, Thomas, and Gardner, William: 'Brighthelmston' (Brighton), 1779. (*Plate 7*.) BL Maps K.Top. XLII.15.
51. Donne, Benjamin: 'A new and correct plan of the City of Bath', 1810. (*36*.) BL Maps K.Top. XXXVII.23.
52. Hamerani, Otto: Bronze medal, 'Appeal against the House of Hanover', 1721. Private collection.
53. Ratzer, Bernard: 'Plan of the city of New York', 1776 (*Plate 8*.) BL Maps K.1. Tab.44, f.27, f.28.
54. Hornor, Thomas: 'Plan of the town and parish of Kingston upon Thames', 1813. (*Plate 9*.) BL Maps K.Top. XL.15–3.
55. 'Plan of the city of Perth'. *From*: Wood, John: *The Town Atlas*, 1823. (*37*.) BL Maps C.21.e.4.
56. Sudder Board of Revenue: 'Map of the Station of Allahabad and adjacent country', 1862. BL Maps 53230.(2).
57. G M Hopkins & Co: *Atlas of the city of Buffalo*, 1872. BL Maps 22.c.38.
58. Heywood, John: 'Pictorial map of Manchester and Salford', 1886. (*39*.) BL Maps 3215.(16).
59. Ordnance Survey: *Plan of Huddersfield*, 1890. (*Plate 10*.) BL Maps O.S.T.199.
60. Anon: Woodcut map of Peking, *c*.1900. (*41*.) BL Maps 30.b.54.
61. Bulman, W J: 'Bird's-eye view of Winnipeg', 1900. (*38*.) BL Maps 70796.(17).
62. Pouillot, Leon: 'Eventail cycliste', *c*.1895. BL Maps C.27.e.24.
63. Bacon, George Washington & Co: 'Bacon's index map of Margate', *c*.1893. BL Maps 3110.(15).
64. Philip, George & Son: Cover from 'Philips' new plan of Greater Liverpool', 1895. BL Maps 3200.(22).
65. Bartholomew, John: 'Plan of the city of Edinburgh, with Leith and suburbs', 1891. (*40*.) BL Maps 19.e.16.
66. Maps of London poverty. *From*: Booth, Charles: *Labour and Life of the People*, 1891. (*Plate 11*.) BL Maps C.21.a.18.
67. Plan of Catalan Bay, Gibraltar. *From*: Chervin, Nicolas: *Documens relatifs à l'épidémie de fiévre jaune*, 1830. BL 1168.h.30.
68. Map of an epidemic in Glasgow. *From*: Perry, Robert: *Facts and observations on the sanitory state of Glasgow*, 1844. (*Plate 12*.) BL C.T.249.18.
69. Map showing the areas served by two London water companies. *From*: Snow, John: *On the mode of communication of Cholera*, 1855. (*42*.) BL 7560.e.67.
70. Anon: 'The siege of Colchester by the Lord Fairfax, as it was with the line and outworks', *c*.1648. BL Maps 2390.(1).
71. De Hooghe, Romeyn: 'Luxemburgum', 1684. (*44*.) BL Maps C.9.e.4.(63).
72. Boyer, Abel: 'A map and plan of the town of Preston', 1715? BL Maps 3230.(1).
73. Meunier, L, and Rouiller, P: 'Paris en Mai 1871'. BL Maps 16114.(8).
74. Leake, John: 'An exact surveigh of the . . . ruines of the City of London', 1669. (*45*.) BL Maps Crace I.50.
75. Goad, Chas E: *Insurance plan of Leeds*, 1886. (*Plate 13*.) BL Maps 145.b.14.